WITHDRAWN
UTSA LIBRARIES

Science, Technology and Taxation

Series on International Taxation

VOLUME 40

Founding Editors

Richard Doernberg, Kees van Raad and Klaus Vogel †

Senior Editor

Professor Dr Kees van Raad,
International Tax Center, University of Leiden

Managing Editors

Professor Ruth Mason,
University of Connecticut School of Law
Professor Dr Ekkehard Reimer,
University of Heidelberg

The titles published in this series are listed at the end of this volume.

Science, Technology and Taxation

Robert F. van Brederode

Law & Business

Published by:
Kluwer Law International
PO Box 316
2400 AH Alphen aan den Rijn
The Netherlands
Website: www.kluwerlaw.com

Sold and distributed in North, Central and South America by:
Aspen Publishers, Inc.
7201 McKinney Circle
Frederick, MD 21704
United States of America
Email: customer.service@aspenpublishers.com

Sold and distributed in all other countries by:
Turpin Distribution Services Ltd
Stratton Business Park
Pegasus Drive, Biggleswade
Bedfordshire SG18 8TQ
United Kingdom
Email: kluwerlaw@turpin-distribution.com

Printed on acid-free paper.

ISBN 978-90-411-3125-6

© 2012 Kluwer Law International BV, The Netherlands

All rights reserved. No part of this publication may be reproduced, stored in a retrieval system, or transmitted in any form or by any means, electronic, mechanical, photocopying, recording, or otherwise, without written permission from the publisher.

Permission to use this content must be obtained from the copyright owner. Please apply to: Permissions Department, Wolters Kluwer Legal, 76 Ninth Avenue, 7th Floor, New York, NY 10011-5201, USA. Email: permissions@kluwerlaw.com

Printed and Bound by CPI Group (UK) Ltd, Croydon, CR0 4YY.

Table of Contents

List of Editors and Authors ix

Introduction 1

PART I
Empirical Evaluation of Taxation 9

CHAPTER 1
Empirical Legal Studies and Taxation in the United States 11
§1.01 Introduction 11
§1.02 A Short Overview of Early Empiricism in Legal Studies 12
§1.03 ELS as an Academic Discipline Sui Generis 17
§1.04 Research Methodology 18
§1.05 ELS Studies in the Area of Taxation 21
§1.06 Summary and Conclusions 49
§1.07 References 50

PART II
Statistical Science in Tax Auditing 55

CHAPTER 2
Statistical Sampling in Tax Compliance Auditing in the USA 57
§2.01 Introduction 57
§2.02 Why Sample? 58
§2.03 What Is Statistical Sampling? 59
§2.04 Sampling Authority and Sampling in Other Audit Fields 62
§2.05 Sampling Theory: Simple Random Sampling Example 66
§2.06 Dealing with Tax Compliance Populations with Large Variances 72
§2.07 Planning the Sample: Population Refinements and Establishing the Frame 75

Table of Contents

§2.08	Projecting the Sample Results	81
§2.09	Stratification	88
§2.10	Problems in Applying Statistical Sampling	92
§2.11	References	103

CHAPTER 3
Legal Issues Surrounding Sampling — 105

§3.01	Introduction		105
§3.02	Permissibility of Sampling		107
	[A]	Burden of Proof	107
	[B]	Type of Tax	108
	[C]	Confidence Interval	111
	[D]	Taxpayer's Acceptance and Sampling Agreements	112
	[E]	Fairness	113
§3.03	Extrapolation		114
§3.04	Interest and Penalties		115
§3.05	Concluding Remarks		116
§3.06	References		118

PART III
Technology Use in Government Tax Administration — 119

CHAPTER 4
Technology and Taxation in Developing Countries — 121

§4.01	Introduction			121
§4.02	Technology and Tax Administration			123
	[A]	*Challenges Facing Tax Administrations*		123
		[1]	Size of agricultural and informal sector	123
		[2]	Use of the financial sector	124
		[3]	Organizational change	125
		[4]	Administrative capacity	127
		[5]	Political will	128
	[B]	Using Technology to Improve Tax Administration		128
		[1]	Tracking taxpayers	130
		[2]	Information reporting and withholding	131
		[3]	Processing returns and payments	133
		[4]	Auditing taxpayers	134
		[5]	Services to taxpayers	135
		[6]	Management of tax administration	136
§4.03	Technology and the Design of Tax Systems			137
	[A]	Tax System Design		137
	[B]	Tax Instruments		139
		[1]	Value-added taxes	139
		[2]	Excises	140

Table of Contents

	[3]	Trade taxes	140
	[4]	Property taxes	141
	[5]	Income taxes	142
	[C]	Tax Reform Process	143
§4.04	Privacy and Privatization – Some Cautionary Notes		143
	[A]	Privacy – Big Brother Is Watching	143
	[B]	Privatizing Tax Administration	147
§4.05	The Future Relationship of Tax and Technology		148
§4.06	References		150

CHAPTER 5
Information Technology in the U.S. Tax Administration 159
§5.01 Introduction 159
§5.02 The Evolution of Information Technology in Tax Administration 160
  [A] Early Systems Focused on Processing Incoming Data and Maintaining Accounts 161
  [B] Efficiency Is Driven by the Automation of Work Processes 161
  [C] Automation and Electronic Filing Drive Significant Processing Cost Reduction 162
  [D] Information Matching Programs Provide Significant Compliance Presence 163
  [E] Web-Based Systems Offer Improved Service to Taxpayers 163
§5.03 Planning for the Future: Developing an IT Program Roadmap 164
§5.04 Technology Challenges Facing Tax Agencies for the Future 166
  [A] Support Delivery of Current Operations 166
  [B] Carry Out New Responsibilities 167
  [C] Protect Sensitive Data 168
  [D] Drive Business Efficiency and Cost Reduction 169
  [E] Improve Service to Taxpayers and Tax Professionals 170
  [F] Address Noncompliance with the Tax Laws 171
§5.05 Conclusion 175
§5.06 References 175

PART IV
Technology Use in Tax Compliance and Tax Planning 177

CHAPTER 6
The Role of Automation in Improving Indirect Tax Reporting and Compliance 179
§6.01 Introduction 179
§6.02 Regulatory Environment 181
§6.03 Technological Solutions 182
  [A] Capabilities and Limitations of Enterprise Resource Planning Systems (ERP) 183

	[B]	Tax Solutions beyond the ERP		186
		[1] Tax Engines		186
		[2] Hosted Tax Solutions		188
		[3] Reporting Solutions		188
§6.04	The Implementation Process			188
§6.05	Benefits and Value of Tax Technology Application			190
§6.06	The Role of Technology in Supporting Tax Strategy			191
§6.07	Conclusions and Predictions			192
§6.08	Bibliography			193

CHAPTER 7
Automation Technology in Special Customs Programs and Preferences — 195

§7.01	Introduction			195
§7.02	Government-Mandated Automation			196
	[A]	Mexico: The IMMEX Program and Automated Inventory Controls		197
		[1] Overview and History		197
		[2] Automated Inventory Controls		197
	[B]	RECOF in Brazil		198
		[1] Overview and History		198
			[a] Benefits	199
		[2] Requirements for Participation		200
		[3] RECOF Software		200
§7.03	Automation Necessary to Realize Benefits			201
	[A]	U.S. Foreign-Trade Zones		201
		[1] Background		201
		[2] Benefits		202
		[3] Inventory Control and Recordkeeping System		203
		[4] Petroleum Refining FTZs		203
	[B]	Customs Warehousing in the European Union		205
		[1] Overview and Benefits		205
		[2] Focus on Automation		206
		[3] Customs Freight Simplified Procedures (CFSP)25		206
	[C]	Processing Trade in China		207
		[1] Overview and History		207
		[2] Primary Requirement for Participation in PTC: Customs Handbook		207
		[3] Development of the Electronic Customs Handbook		209
§7.04	Use of Automated Tools			210
§7.05	Conclusion			211
§7.06	References			212

List of Editors and Authors

Dr. Richard M. Bird (Ph.D., Columbia, 1961; M.A., Columbia, 1959; BA, King's College, 1958) is Professor Emeritus of Economics, Rotman School of Management, and Adjunct Professor and Associate of the Institute for Municipal Governance and Finance, Munk School of Global Affairs, University of Toronto. He is also Distinguished Visiting Professor, Andrew Young School of Public Policy, Georgia State University, and Adjunct Professor of the Australian School of Taxation and Business Law, University of New South Wales. He has served with the Fiscal Affairs Department of the IMF, and has been a visiting professor in the United States, the Netherlands, Australia, Japan, and India and a frequent consultant to the World Bank and other national and international organizations. He has published extensively on tax and public finance issues, with special emphasis on the fiscal problems of developing and transitional countries.

Dr. Robert F. van Brederode (Ph.D., Amsterdam, 1993; LLM, Utrecht, 1985) is a tax lawyer concentrating in the area of VAT and other transaction taxes. He has over 26 years experience in the field practicing on both sides of the Atlantic, currently in solo practice. In addition, he teaches courses as Adjunct Professor at New York University, School of Law, Graduate Tax Program. Previously, Dr. Van Brederode was a partner in PwC, leading the Netherlands VAT & Customs practice, and Professor of tax law at the Erasmus University, School of Economics. He is the author of *Systems of General Sales Taxation: Theory, Policy and Practice* (Series on International Taxation, No. 33, 2009), and editor of *Immoveable Property under VAT: A Comparative Global Analysis* (Series on International Taxation, No. 37, 2011). He may be reached at robvanbrederode@aol.com.

Andrew T. Buckler served for 32 years at the IRS where he co-led the development of the IRS Modernization Vision & Strategy, and later served as Director of the agency's largest systems modernization program, the Customer Account Data Engine 2. His final assignment was as Special Advisor to the Deputy Commissioner Services & Enforcement for information technology issues. Andy retired from government service in 2010 and is currently a director in the Federal Services practice of Deloitte Consulting LLP, Washington, DC.

List of Editors and Authors

Erik van der Hoeven is the head of the Global Tax Technology Group with Alvarez & Marsal Taxand, LLC in the United States. He specializes in system solutions that lower the total cost of tax compliance through the centralized management of transaction taxes, such as Sales and Use Taxes, VAT, environmental taxes and Customs duties. He has worked extensively with global ERP systems, such as SAP and Oracle as well as legacy systems. Previously, Mr. Van der Hoeven served as senior-manager at Ernst & Young in Amsterdam and Chicago, and as Director of Business Consulting at Sabrix (acquired by Thomson Reuters in 2009).

Stephen W. James is a partner in KPMG's U.S. Indirect Tax practice based in Mountain View, California. Mr. James specializes in the deployment of tax technology to support global VAT and similar taxes and to manage and facilitate global indirect tax determination, calculation, compliance and reporting. Prior to his career in professional practice, Mr. James served as an Officer of Her Majesty's Revenue and Customs in the United Kingdom.

Dr. Robert W. McGee is a Professor of Accounting at Fayetteville State University in North Carolina. He is an attorney and CPA and has published more than 50 books and 600 articles and book chapters. He has worked or lectured in more than 30 countries in the former Soviet Union, Europe, Asia, Africa, Latin America and the Caribbean. He drafted the accounting law for Armenia and Bosnia and reviewed the accounting law for Mozambique. Several studies have ranked him #1 for business ethics scholarship.

William M. Methenitis is the Global Director of Customs and International Trade for Ernst & Young. He is a Phi Beta Kappa graduate of the University of Texas (with Highest Honors), and holds a juris doctorate from the Southern Methodist University School of Law (cum laude, Order of the Coif). He may be reached at william.methenitis@ey.com.

Robert Schauer works for the Multistate Tax Commission (MTC) since 2006 as a Computer Audit Specialist and as an instructor for various training classes offered by the MTC, including training on statistical sampling. At the MTC, he developed software and materials that are used in training and audits. Mr. Schauer was employed for 24 years with the Washington State Department of Revenue (WADOR). During his time with WADOR, he served as an auditor and then later as a computer audit specialist. He was a key individual in the development of the sampling program and policy currently used by the WADOR. Mr. Schauer has a degree in accounting from the University of Portland, and has co-authored several articles on statistical sampling that have been published in the Journal of Government Financial Management.

Eric M. Zolt (B.S., Wharton School, University of Pennsylvania, 1974; M.B.A., University of Chicago, 1975; J.D. University of Chicago, 1978) is the Michael H. Schill Professor of Law, UCLA School of Law. Before joining UCLA, he was a partner in the Chicago law firm of Kirkland & Ellis, where he specialized in individual and corporate tax matters. Before practicing law, Eric was on the research staff of the Center for Policy Alternatives, Massachusetts Institute of Technology. While on leave from UCLA, he served in the U.S. Department of the Treasury from 1989 through 1992, first as Deputy Tax Legislative Counsel in the Office of Tax Policy and later as the Director of Treasury's Tax Advisory Program for Eastern Europe and the Former Soviet Union. Eric serves as a consultant to the Treasury Department, US AID, the World Bank and the International Monetary Fund and he has provided tax policy advice in over 30 countries. In 2002, Eric co-founded and served as the first Chair of the Executive Committee of the African Tax Institute, a training and research facility for tax policy and tax administration located at the University of Pretoria. Eric has been a visiting professor at Harvard Law School, the University of Toronto, Yale Law School, and at University of Aix-en-Provence.

Introduction

Robert van Brederode

Technological and scientific developments and improvements have changed society for millennia from the invention of the wheel, agriculture to the computer and Internet. The automobile put the horse out of a job, satellites have improved weather forecasting and global communication, airplanes have transformed most of us into globetrotters. This book is about the interaction between technology and science on the one hand and taxation on the other. The Internet, for example, has enabled the sale of digitalized products and that change had direct repercussions on sales taxes that piggyback economic transactions. Where tax is levied on the sale of a physical product, digitalized delivery may escape taxation providing a competitive edge to virtual shops over brick-and-mortar shops, jeopardizing the principle of economic neutrality which is the cornerstone of equal treatment under value added tax (VAT) and other general sales taxes. Another example is the use of tax incentives to encourage the development of certain new technologies. On the flip side of the coin, the manufacturing of demerit products or the use of certain production methods may be discouraged by (extra) taxation. The growing utilization of so-called green taxes to steer businesses to the application of more environmentally friendly production methodologies are a striking modern example of this.

However, the intention of this book is not to provide a historical overview or even a complete picture of all occurrences where taxation exercises influence on or is influenced by technology and science. This book only provides a selection of areas where taxation interacts with science and technology, divided in four parts. Part I focuses on empirical evaluation of tax legislation; Part II is dedicated on the use of statistical science in the field of tax auditing; Part III is directed towards technology use in government tax administration; and Part IV concentrates on technology use by businesses for the purpose of improving tax compliance and tax planning.

It starts (Chapter 1) with reflection on the efficiency of tax legislation by presenting an overview of empirical studies conducted in the area of taxation in the context of Empirical Legal Studies (ELS). ELS are a relatively new phenomenon, starting as a movement in the mid-1990s but with its roots tracing back to the early 20th century. Because the law touches almost every aspect of life and society, ELS cover a broad field of study, but in the context of this book we are interested in empirical research in the field of taxation. The chapter explains what ELS is and how it has developed over the years to its present state as a recognized and respected academic discipline. It includes an overview of types of research conducted in the earlier years. ELS uses statistical and other methodology developed in the social sciences to study the impact and effectiveness of legislation and policies. In other words, are stated policy and legislative objectives indeed realized? For example, some studies focus on determining whether higher taxes on tobacco and alcoholic products realize the anticipated decrease in consumption. Empirical legal studies can assist law makers in making better legislation and in repairing failing laws and policies. They may also help with the realization of the relative effectiveness of laws and regulations. There clearly are limitations as to what can be achieved through legislation and regulation, and ELS provide a perfect reality check.

Tax auditing is the focal point of the second part of the book. Especially in respect of larger companies, it is unrealistic and impractical to conduct a full audit, i.e., a complete review of each book entry and every invoice received and issued. Neither the tax administration nor the taxpayer prefers a full audit because of the restraint on limited human resources and the time investment required on both sides to go through the full audit process.

Chapter 2 explains what statistical sampling is, what choice of methodology exists, and how to conduct a sampling audit. The basic definition of a sample is any part of the population selected for audit inspection. A population is an aggregation of business transactions or bookkeeping entries. Usually, in tax compliance auditing, populations consist of economic transactions that are of audit interest or elements that make up the tax return. The population, or accounting, may be available to the auditor in either paper or electronic forms. In tax compliance auditing, all different kinds of sample designs are used, which are discussed and evaluated in this chapter. But regardless of the design, there are only a few basic objectives for using a sample in an audit, including discovery of errors and determination of their frequency, and estimation of the total tax error. The total tax error may be a net underpayment (tax deficiency owed to the government), net overpayment (where a refund may be in order), or net of zero (a taxpayer who is in compliance). When the auditor wishes to estimate a total of some sort, typically the auditor will extrapolate or project the sample results to represent a total that would have resulted from a complete audit of the entire population. Logically, the objectives for sampling are actually the same as that for performing the audit itself. In estimation sampling, where a total value is projected from the sample results, it is used to represent the total that would have resulted had a complete examination been made presuming that the same general audit techniques were used. However, the sample estimate is likely different than the true total. This difference between the projected total and the true value being estimated is called the sampling error. Depending on the nature of the population, the sampling design, the size of the sample, and indeed what sample is drawn by chance, the degree of sampling error may be small or large. Sophisticated sampling procedures can be used to both quantify and control this error. Accuracy is key to the reliability of statistical sampling and in this context the author explains in detail the importance and determination of

Introduction

both one- and two-sided confidence interval. The author details potential pitfalls that may render the sampling less reliable or even insufficient as a basis for a tax assessment. Sampling methods and practices are illustrated with numerical examples. In his final remarks, the author explains practical problems in applying statistical sampling in a tax audit.

Chapter 3 approaches sampling in tax audits from a legal perspective and discusses whether and, if so, under what circumstances statistical sampling is acceptable as proof of underpayment and as a basis for penalties, because sampling does not constitute an exact determination but provides only an estimate of taxes due. It discusses whether the permissibility of sampling would depend on the allocation of the burden of proof to either the tax payer or the taxing authority. In the author's opinion sampling should be allowed regardless of whether the burden of proof rests with the taxing authority or the taxpayer. However, the allocation of proof has an influence on the amount of tax assessment. Where the burden of proof rests with the state, the assessment should be based on the minimum error position given the width of the confidence interval. That would be fair, because that is all the state has actually proven to be wrong. On the other hand, if the burden of proof rest with the taxpayer, a tax assessment calculation based on the mid-range of the confidence interval (point estimate) appears to be fair and appropriate. Interest charges should be calculated on the same basis. However, the same rationale does not apply to penalty charges. Penalties will vary depending on the degree of culpability, which cannot satisfactory be assessed through sampling.

Another point of discussion is whether the type of tax is relevant in determining the use of sampling, this on the basis of some rulings by the Netherlands Supreme Court. The conclusion is that type of tax is irrelevant even when a business is used as collector for taxes economically resting on someone else, such as wage taxes and VATs, because the collector has a direct liability for these taxes.

The use of sampling agreements is recommended as a tool to improve understanding of the process by the taxpayer. Goal of a tax compliance audit is a true and accurate assessment. Allowing sampling audits seems to be both rationally consistent and practical. However, where it can be shown that the auditor used poor methods, it would seem that a taxpayer could rebut an assessment. That should also hold when a taxpayer is party to a sampling agreement with the taxing authority.

Interesting is the author's opposition to extrapolation as found in one of the cited court cases. A sample has only meaning for the population from which it has been taken and cannot reasonably provide guidance in terms of error for populations that are not included in the sampling audit. Even is the same type of error is made in other years, the frequency of the error may vary from year to year.

Another important topic of the book is the use of technology in tax management (by the revenue services) and tax compliance management (by corporate taxpayers). Government tax management, Part III of the book, is covered in two chapters.

Chapter 4 gives a general overview of the complexities encountered by the introduction of technology in tax administration in developing countries. It comes as no surprise that technology has also affected how tax systems are designed and administered in developing countries. These changes have not always been for the better. Technological change continues. Most countries have now moved from rooms full of clerks posting entries by hand in large ledger books to widespread use of computers to administer their tax systems. This transition has been incomplete and uneven in developing countries. Major differences exist among and within developing countries, both with respect to how their tax systems are designed and administered, and, more generally, with respect to how technological advances have changed the manner in which their economies operate. The authors conclude that technology is not a 'magic bullet' to solve the manifold problems of development taxation, but it may provide part of the answer for many countries. One important question is what the necessary preconditions in developing countries are for successful adoptions of new technology. The main focus of this chapter is on how (1) technology may be used to improve tax administration, and (2) technology may potentially change the design of tax systems in developing countries.

The economic environment, especially the size of the agricultural and informal sectors, is crucial for the administration of taxes in developing countries. These sectors are relatively large in developing countries and this reduces the tax base that tax administrations can reach. A related issue is the interaction between the formal and informal sectors. By providing additional tools to observe and monitor transactions and taxpayers, technology may significantly enhance the ability of tax authorities to detect economic activity in the informal sector. The growth of the financial sector also broadens the potential scope of taxation and makes administration of certain taxes easier down the supply chain. From several studies it is clear that improvements in tax administration require changes in organization and methods. Information technology can help facilitating such a transformation.

Another problem found in developing countries is the scarcity of administrative capabilities in terms of skilled human capital. This may hinder technological advancement to some degree but, on the other hand, technological innovation may also be able to substitute for inadequate human capacity. The authors provide an overview of several roles that tax administrators play and how technology can help enhance performance in that regard.

From an administrative perspective, most taxes collected in developing countries come from a relatively small number of tax collecting agents. Accurate tracking of fiscal flows through such large entities, which probably account for 80% or more of current collections in many countries, is critical to successful tax administration. Tight control over payments and liabilities of large taxpayers is critical for success and the creation of large taxpayer units has been successful. Given the importance of the VAT as the primary source of revenue in many developing countries, the VAT is likely the tax that has the greatest potential for substantial gains from the use of technology. Key in enforcing VAT compliance is cross-checking purchases of one taxpayer against sales recorded by others, but few developing countries have systematically used such programs to detect under-reporting.

Tax systems vary greatly in their reliance on different tax instruments to generate revenue in order to support government operations. Developing countries rely much more on trade taxes, excise taxes, and general consumption taxes, and less on income tax, than developed countries. Technology may change this calculus of tax structure.

Introduction

Also, it may enable tax authorities to change the policy mix towards taxes that can now be more effectively administered. The authors discuss the specifics of these taxes as they relate to developing countries. In particular for trade related taxes there are several technological improvements available to better monitor the movement and values of goods entering a country, such as Tamper-resistant Embedded Controllers (TRECs), Radio-frequency ID (RFIDs), and Laser Surface Authentication (LSA).

Good tax administration relies on the good use of information. Good tax reform similarly depends on the good use of information. Improvements in information technology allow countries to administer existing taxes better, and to change both the administration and the structure of taxes to better achieve their developmental objectives. Techniques for developing tax forecasting and tax simulation models are now advanced enough that even unsophisticated tax departments can use such models to provide information to policymakers.

Last but not least, the authors discuss privacy concerns raised by the progress in information technology. This is a vital subject for policymakers and individual citizens alike. Conceptions of rights to privacy differ among societies and, therefore, the need for privacy protections will be defined differently among nations and cultures. The real danger from new technology may not be the erosion of the tax base as taxpayers use technology to avoid tax, but the erosion of privacy as governments take defensive action to protect revenue in the emerging digital economy. The authors suggest that the only viable answer is for citizens to reach a consensus that permits access to private activities necessary for the sustenance of the public community, without allowing such information to be misused. Of course, technology does also provide safeguards to prevent unauthorized use of personal information. However, the more nuanced and complex taxes become, the more personal information will be needed to assess whether a taxpayer has fully complied and whether deductions and other incentives are rightfully claimed. An opposite approach–they do not mention — would be to choose and design taxes that do not require such intrusive information to ensure compliance, e.g., single-rated consumption taxes and proportional (flat) income taxes. In that case, of course, what is gained in privacy protection will be lost in terms of tax policy manoeuvrability.

In contrast, Chapter 5 describes the experience with technology in the tax administration of a single developed country, the United States. The experience in the United States demonstrates the Herculean effort required to adequately process and manage massive amounts of data. Tax administration is primarily an information management business. Information is received in the form of returns, payments, information documents and many other types of submissions, and that information is supplied by many entities, including individual and business taxpayers, tax professionals, financial institutions, payroll servicing firms and other government agencies at the federal, state and local levels. In short, it is a high volume and very complex information environment. For example, the IRS has to process 230 million tax returns and 2.7 billion third-party information documents, running a wide variety of hardware platforms used by over 94,000 IRS employees. With the explosion of IT capabilities in all aspects of modern life, tax agencies face many new challenges. The increasing speed, complexity and global scope of economic transactions challenge the ability of tax agencies to understand and monitor activity for tax implications. In addition, there are those who promote tax evasion schemes via off-shore financial institutions and structures. Tax agencies are challenged to obtain and analyze information to help them to identify and address such schemes. The expectations of the public have changed with regard to the speed and manner in which services are provided. Add to these complexities, the demand for change in IT systems as a result of tax law changes, and new responsibilities, not directly related to taxation.

This chapter gives an overview of the current state of the IT environment in the context of the long term evolution life cycle. IT systems have been developed over time and many of the early systems have remained in production. As a result, the IT environment includes systems that were constructed at different times, using different technology platforms, different programming languages and providing different types of functionality. This leads to a very complex operational environment, and sometimes limits the capabilities that can be added to existing systems. The technological foundation was built around account maintenance, the next wave of automation focused on providing automated tools to IRS employees in the customer service and compliance functions. Electronic submission of third party information documents was the next step in the IT evolution, enhancing efficiencies and reducing administrative errors both on the side of the tax administration and on the side of taxpayers. New developments include electronic filing, ability of taxpayers to track refunds on line, and website-based provision of information. The IRS and other tax agencies are resource limited, both from a funding perspective and in terms of their internal capacity to manage change. As a result, tough decisions must be made to prioritize the allocation of IT resources to assure that the key strategic objectives of the agency are being addressed. Effective IT planning requires partnerships among agency business leadership and external stakeholders, including the executive branch, legislative oversight, tax industry representatives and taxpayer groups. Another vital task of the tax agency is protection of taxpayer personal data, which is also critical in maintaining a high level of voluntary compliance. Tax data is subject to very strict disclosure rules, and IRS is working hard to assure the protection of its data. However, the challenges associated with systems security and data protection continue to grow for tax agencies. All government agencies, including the revenue service are under pressure to control and reduce costs. From a technology perspective, cost control has two aspects: controlling the cost of IT operations; and providing new capabilities to enable agency business functions to improve their productivity. One very important opportunity for both service improvement and cost reduction is through the implementation of online self-help service to taxpayers.

A separate challenge is offered by non-compliance which is a multi-dimensional problem. The difference between tax legally due and tax actually reported has three sources: Non-filing, underreporting and underpayment. To address non-compliance within each type of tax and population segment, IRS must follow a series of fundamental steps, each of which is supported by technology.

The author concludes that the rapid evolution of information technology will continue to be a double-edged sword, posing both major opportunities and significant challenges. Opportunities are found in improvement in efficiency and effectiveness,

Introduction

the offering of new services and enhanced capabilities in the area of compliance enforcement. Challenges are offered by growing demand for IT capabilities and the requirement to develop and deploy new technology capabilities with flat or even shrinking budgets.

The taxpayers' perspective as to the use of technology is presented in Part IV, also through two chapters.

As a more general overview, Chapter 6 explain the ins and outs of the use of tax engines to improve the compliance rate in regard of indirect tax, especially VAT and retail sales tax. The global expansion of businesses and the concurrent shift to indirect tax requires companies to comply with increasingly complex regulations and to report correctly their tax positions in a multitude of jurisdictions. The authors explain the regulatory environment and the pressure it puts on stricter internal control and improved tax compliance. The challenges faced by businesses are, to a large extent, systemic and process driven and, therefore, they can be addressed through technology. I essence, it comes down to leveraging existing business data and reduce the impact of the human element to reduce costs, effort and risk. The Enterprise Resource Planning (ERP) application fulfill a central role as it enables most of the business functions that have tax consequences, such as sales, procurement, logistics, and inventory. However, generally, ERP systems offer little in tax functionality or tax content. Also, the tax function within an ERP system relies heavily on IT and other nontax resources to implement change, which, in itself, enhances the risk of omissions and mistakes. The authors subsequently discuss the tax functionality of both SAP and Oracle, the most commonly used ERP systems among larger companies, with a focus on VAT functionality. Both provide a framework for tax but no content which, therefore, needs to be created and maintained. There exist a number of tax solutions and services that seek to replace the native functionality in the ERP with enhanced capabilities as well as content such as tax rates, rules and tax logic. Leading bold-on tax engines, such as Sabrix, Taxware and Vertex, are discussed. In essence, these engines take data elements that describe a transaction and use those data elements to make a tax decision in much the way that a tax professional would interpret the facts of a transaction when determining its treatment. The engines will drive tax decisions that are more easily managed, archived and audited and, importantly, will reduce the influence of users on the final tax outcome. It is crucial to understand that these engines need to be integrated with the ERP system to ensure that data from the latter will be passed to the former. The authors also explain hosted tax solutions, offering access to a third-party tax engine (outsourcing), and reporting solutions, which are software packages for tax return preparation, either focused on a particular or on multiple jurisdictions. A detailed description is offered of the multi-phase tax technology implementation process. The complexity of these processes cannot be overestimated and the authors identify a number of associated issues that businesses should be aware of. Tax technology offers a number of advantages. An interesting one is the ability to support the tax department through modeling and testing anticipated scenarios to predict the outcome for tax and the impact of changes.

The authors predict an evolution of tax technology rather than a revolution driven by the compliance demands inherent to complex tax systems, such as the one of

Brazil. Looking forward, they expect enhanced focus of tax administrations on business systems review where the accuracy of systems is more and more becoming the determinant factor for measuring the rate of tax compliance. And systems will acquire a much higher level of integration driven by market demand for technologies that can work together. Faster communication between tax payers and tax administration through electronic reporting and revenue transmission will become a growing global practice.

Chapter 7 is focused on a single field of tax law and explains technology use in the area of Customs procedures. Customs automation is often thought of as compliance focused, keeping necessary data for customs declarations and facilitating electronic filing and recordkeeping. Automation, however, has also played a big role in tariff reduction planning — tax planning for customs duties — in the two areas most critical to global traders, special programs and trade preferences. Some of this automation is government mandated, essentially an entry fee for accessing tax savings, and in this context the IMMEX program in Mexico and the RECOF in Brazil are explained and discussed. More frequently, automation enables the tax savings, whether in the form of a comprehensive automated accounting system for a special program, or as a tool allowing rapid assessment and trade preference scenario planning. In this context US foreign trade zones, EU Customs warehouse regimes and Processing Trade in China are explained. In a separate paragraph, the author discusses the complexities of automating origin determination. Looking ahead, as business looks to squeeze additional costs out of the supply chain using special customs programs and trade preferences, and as government enforcement focuses on the compliance requirements for special program and trade preference eligibility, the author predicts that automated systems will play an even more critical role in assisting business to achieve savings and meet compliance obligations.

Although this book is far from complete, we believe that the selection of subjects covered demonstrate how multi-faced the interaction between science, technology and taxation is. The scope of this interaction and its complexity will only grow. This book is current until March 2011. Because the chapters in this book are rather diverse in subject, we have decided not to offer a consolidated bibliography but to provide bibliography for each chapter separately.

Part I
Empirical Evaluation of Taxation

CHAPTER 1
Empirical Legal Studies and Taxation in the United States

Robert W. McGee and Robert F. van Brederode

§1.01 INTRODUCTION

Both traditionally and historically, legal scholarship has been largely based on doctrine and theory. When we went to law school, some 30 years ago, a law student was challenged to take a class in legal philosophy, sociology of law, or law and economics. Apart from legal philosophy which was held in intellectual esteem, the behavioral studies were only minor classes, operating at the periphery of the curriculum. Law students felt more acquainted with Locke, Hobbes and Montesquieu than affiliated with Durkheim or Levi Strauss who generated name recognition at best. Nevertheless, both sociology of law and law and economics have a proud academic tradition with roots going back to the early 20th century. Empirical Legal Studies (ELS) is a relatively new phenomenon, starting as a movement in the mid-1990s, although these types of studies have formed the backbone of the law and society tradition since its inception in the 1960s. ELS became the talk of the town — so to speak. Harvard was first in opening an ELS program in 1996, and in 2004 the *Journal of Empirical Legal Studies*[1] was launched. Also, there is now an official ranking of top ELS law schools and since 2006 an *Annual Conference on Empirical Legal Studies*. For those continuously interested in the field, there is also a rather active Empirical Legal Studies blog.[2]

So what is ELS and what do ELS researchers exactly do? What kind of research tools and methodology do ELS researchers apply? How is it different from sociology of law and law and economics, if at all? These questions are addressed in section 1.03 of this chapter in which we will make an effort to define and classify ELS as a field of study

---

1. *Journal of Empirical Legal Studies* (Blackwell Publishing, first published in 2004).
2. http://www.elsblog.org (last visited January 15, 2011).

and assess its general value. ELS may now be considered as part of the mainstream within the legal academy, it did not arise out of the blue of course but finds its roots in earlier empirical scholarly work, and we believe it is relevant to discover the historical perspective in our discussion of the present status of empiricism in the field of legal studies (section 1.02) because it allows us to make a comparison between the empirical research done today with its predecessors.

It will come as no surprise that empirical legal studies cover a broad field of study, since the law itself touches almost every aspect of life and society. A joint ELS bibliographical project by UCLA and Cornell law school shows as many as 380 different research topics, varying from affirmative action, asylum, attorney and client, computer software, mental state, gun control, scientific evidence, hate crimes, illegal aliens, church and state to capital punishment, compensatory damages and judicial activism.[3] In the context of this book we are interested in empirical research in the field of taxation. The bibliography offers 5 sub-headings on tax for a total of 51 studies. In section 1.05 of this chapter we will provide an overview, explaining subject matter and conclusions, of each of these studies and its contribution to our understanding of the effective impact of tax law on society relative to its stated goals and objectives. It is important to note that this bibliography is far from complete for two reasons. First, its primary focus is on publications made in the United States, thus excluding foreign research, and second it only lists research in specialized ELS publications, thus excluding studies published in general law reviews or economic journals. Nevertheless, it provides us with some general insights as to the type of subjects, themes and issues in the area of taxation that have triggered the interest of empirical researchers. In addition, we present an overview of empirical research by the first author of this chapter in the field of taxation.

§1.02 A SHORT OVERVIEW OF EARLY EMPIRICISM IN LEGAL STUDIES

Empirical legal studies may have grown into a separate academic discipline as we will discuss in section 1.03 of this chapter, but it traces its origins to the legal empiricism movement of the early 20th century. It has taken a long time for empirical legal research to become recognized as a serious genre. Heise describes it as a path "littered with the carcasses of earlier failed starts."[4]

ELS is the most recent development in legal empiricism and builds on the research conducted by earlier "empirical traditions," such as the sociology and law and law and economics, with which it co-exists and to which it is complementary.[5]

---

3. https://apps.law.ucla.edu/els/ (last visited January 15, 2011).
4. Michael Heise (2002), "The Past, Present, and Future of Empirical Legal Scholarship: Judicial Decision Making and the New Empiricism," *U. Ill. L. Rev* 819, 820.
5. Theodore Eisenberg, "Origins, Nature, and Promise of Empirical Legal Studies and a Response to Concerns," *Cornell Legal Studies Research Paper*, December 17, 2010, at 6.

Chapter 1: Empirical Legal Studies and Taxation in the United States §1.02

In the 1920s and 1930s a wide range of empirically oriented research on law was produced, including topics such as banking, bankruptcy, criminal courts, debt, divorce, judicial staffing and judicial selection, juries (both petit and grand), jurisdiction, legal education, legal needs and legal aid, legal profession, litigation, procedure, and small claims. Almost all of the research was done by American scholars focusing on the situation in the United States, which comes as no surprise since the fast majority of early research was linked to the legal realist movement which was essentially an American movement.[6]

Criminal law or criminal justice was the focal point of much of the earlier studies of the 19th and beginning of the 20th centuries. At first, research was directed to find the causes of crime; then attention switched to the handling of crime by the legal system. In the United States, it developed into the American Institute of Criminal Law and Criminology (1909) and the publication of what is now known as the Journal of Criminal Law and Criminology (1910). One early study focused on the duration of criminal cases in Wisconsin,[7] another addressed the pressing issue of the need to provide public defenders, finding that those represented by public defenders pleaded guilty at a much higher percentage (70) than those represented by private attorneys (49).[8] A third study examined the behavior of judges in minor criminal cases in New York City and found substantial variation in sentencing for similar offenses.[9] How criminal courts handled cases was the subject of a large body of empirical research in the 1920s and 1930s. These studies were focused on descriptive statistics, providing summaries of data on sentencing patterns with little effort to explain these patterns.[10]

Early studies are not limited to criminal law. Civil justice was the subject of a 1920s study comparing the time needed from filing to judgment of several courts in California.[11] Many other studies focused on the duration of litigation, including one that compared trial duration of three levels of courts in Texas over the extension of 30 years using an annual sample of 50-150 cases.[12] Interestingly, the study was geared towards establishing facts and did not try — by the explicit statement of the authors — to make a root cause analysis to explain differences in trial duration. Many studies in the field of civil litigation were conducted, initially categorizing cases to determine relative case load per field of law, e.g., divorce law, negligence, debt and foreclosure, etc. Research also determined what proportion of case per category involved actual court

---

6. Herbert M. Kritzer, "Empirical Legal Studies before 1940: A Bibliographic Essay," *Legal Studies Research Paper Series, Research Paper 09-27*, University of Minnesota Law School, August 2009, p. 2. Also published in *J. Empirical Legal Stud.* 6: 925.
7. Oliver Rundell, "The Time Element in Criminal Prosecutions in Wisconsin," *Bulletin of the University of Wisconsin* 512 (1912).
8. Walton J. Woods, "Necessity of Public Defender Established by Statistics," *Journal of the American Institute of Criminal Law and Criminology* (1916): 230-244.
9. George Everson, "The Human Element in Justice," *Journal of the American Institute of Criminal Law and Criminology* 10 (1919): 90-99. In today's terminology, this is referred to as the problem of sentencing disparity, often countered through systems of sentencing guidelines and mandatory sentencing.
10. Kritzer, *supra* note 6, 925, 932.
11. Sam Bass Warner (1920), "Procedural Delays in California," *California Law Review* 8: 369-383.
12. Robert Slayton and Philip Brown (1939), "A Study of Pendency in Texas Civil Litigation," *Texas Law Review* 18: 1-26.

action, how many as a percentage of total cases were contested, how many decided for the plaintiff or the defended, how many cases were actually appealed and how many of those produced reversals.[13]

Several specialized areas received significant attention from empirically interested scholars in the period preceding the Second World War: Auto accident and related compensation, divorce, small claims courts, evidence, juries, debt & bankruptcy, banking, appellate courts, legal needs, the legal profession, legal education, and the selection of judges.

For example, in 1932 the first study of the compensation process in automobile accidents was conducted,[14] collecting data on 8,849 accidents. Main conclusion was that compensation as a percentage of loss declined as loss increased.[15] Compensation studies are still conducted today, although not limited to auto accidents.[16]

The early studies in the field of divorce were mainly based on court records and collected information on the length of marriages, number of children, grounds for divorce, initiator of the proceedings, duration of the process and financial relief sought. Interestingly, all studies found that most divorces were consensual even if the formal legal framework did not envision it as such and the suggestion was made to adjust legal procedures to accept that reality.[17]

The small claims courts studies focused mainly on the outcome of the cases, i.e. in whose favor (plaintiff or defendant) the case was decided.[18]

An early experimental study surveyed the quality of evidence rules by asking categories of respondents to rank them on a scale from bad to good.[19] This was probably one of the first studies to utilize statistical methods as a foundation for legal research, basically correlations and in essence a t-test for comparing means.[20]

Generally, the diminishing role of juries in civil cases was the key element of jury research. Another issue was and is how juries decide cases compared to judges. The

---

13. See for the dominant study in the late 1930s, Charles E. Clark and Harry Shulman (1937), *A Study of Law Administration in Connecticut: A Report of an Investigation of the Activities of Certain Trial Courts of the State*. New Haven: Yale University Press.
14. *Report, by the Committee to Study Compensation for Automobile Accidents, to the Columbia University Council for research in Social Studies*. Philadelphia: Press of International Printing (1932).
15. *Supra* note 14, p. 67, 91.
16. For example, Kenneth S. Abrahams (2008), *The Liability Century: Insurance and Tort Law from the Progressive Era to 9/11*. Cambridge: Harvard University Press.
17. Kritzer, *supra* note 6, at 18. See, N.P. Feinsinger (1932), "Observations on Judicial Administration of Divorce Law in Wisconsin," *Wisconsin Law Review* 9: 26-48; Leon C. Marshall et al., (1932), *The Divorce Court Volume One: Maryland*. Baltimore: John Hopkins Press; and Leon C. Marshall et al. (1933), *The Divorce Court Volume Two: Ohio*. Baltimore: John Hopkins Press.
18. Kritzer, *supra* note 6, at 19.
19. Steuart Henderson Britt (1940), "The Rules of Evidence — An Empirical Study in Psychology and Law," *Cornell Law Quarterly* 25: 556-580.
20. The t-test is probably the most commonly used Statistical Data Analysis procedure for hypothesis testing. Actually, there are several kinds of t-tests, but the most common is the "two-sample t-test." If sample data shows a different average scores for two groups, the t-test is used to answer whether this represents a real difference or just a chance difference in the samples. By convention, if there is less than 5% chance of getting the observed differences by chance, we reject the null hypothesis and say we found a statistically significant difference between the two groups.

first research in this area from 1934 found little difference in the probability between a jury and a judge finding for the plaintiff in negligence cases. However, in debt cases and breach of contract cases judges appeared to be slightly more favorable, by a margin of 8%, to plaintiffs.[21] This issue has proven to be of lasting interest and it is still studied by contemporary scholars.[22]

Bankruptcy was, of course, a popular research topic in the era following the Great Depression. Several studies were conducted focusing on several aspects related to bankruptcy, including types of businesses involved in bankruptcy, the nature of the debts and assets, the time needed to go through the bankruptcy process, the costs of bankruptcy proceedings and the causes of bankruptcy. As to the latter, it was found that the root cause in many instances was over expansion of consumer credit.[23]

A large study was undertaken as part of the legal realist empirical research enterprise at Yale Law School into the banking practices, especially how various transactions were handled and the extent of conformity to governing legal statutes and legal decisions.[24]

An important and well-known early empirical study[25] centers on the evolution of the jurisdiction of the US Supreme Court, in particular "how the caseload, both volume and content, shifted along with the jurisdictional changes as both the law and the country as a whole evolved."[26] The first publication of 1925 was followed by annual reports, still continuing to this day, all published in the Harvard Law Review. The older research focused on sources of cases, disposition, subject matter of both petitions for certiorari, and decisions on the merits. The contemporary studies, in contrast, focus more on decisional patterns of individual judges. Empirical research was not limited to federal appellate courts; reports were published on the appellate courts at state level as well, notably the Wisconsin Supreme Court,[27] and a study covering all levels of appeal courts in Ohio.[28] The latter covered a limited time period and looked at the basis of jurisdiction, duration of the process and disposition. The broadest study conducted was an analysis of five years of the Maryland Court of Appeals, distinguishing the type of issues at stake, such as criminal, negligence and estates.[29] Another important early study was geared towards determining the judicial influence of each state's Supreme Courts through a survey among law professors, citations in case-books, and citations of

---

21. Charles E. Clark and Harry Shulman (1934), "Jury Trial in Civil Cases — A Study in Judicial Administration," *Yale Law Journal* 43: 867-885.
22. For a more recent study, see Theodore Eisenberg et al. (2006), "Juries, Judges and Punitive Damages: Empirical Analyses Using the Civil Justice Survey of State Courts 1992, 1996, and 2001 Data," *Journal of Empirical Legal Studies* 3: 263-295.
23. See Kritzer, *supra* note 6, at 946-949 and references mentioned there.
24. See Kritzer, *supra* note 6, at 950 and references mentioned there.
25. Felix Frankfurter and James M. Landis (1925), "The Business of the Supreme Court of the United States. A Study in the Federal Judicial System," *Harvard Law Review* 38: 1005-1059.
26. See Kritzer, *supra* note 6, at 950.
27. Richard V. Campbell (1933), "Statistical Survey," *Wisconsin Law Review* 9: 5-10.
28. Silas A. Harris (1933), *Appellate Courts and Appellate Procedure in Ohio*. Baltimore: John Hopkins Press.
29. Herbert M. Brune Jr. and John S. Strahorn (1940), The Court of Appeals of Maryland, a Five Year Case Study, *Maryland Law Review* 4: 343-389.

opinions by other state Supreme Courts and the US Supreme Court to arrive at a prestige ranking list.[30]

The legal needs, especially of low income individuals, form another modern topic with roots in early empirical legal research. The oldest studies focus on the role of public defenders in criminal cases[31] and other legal aid organizations.[32] More modern in methodology is a study to the legal needs, distinguished by specific justiciable problem, using a survey of residents and businesses. The researchers tried to establish the likelihood of professional advice being sought and whether there was a difference in the use of consultants between higher and lower income groups and between individuals and businesses.[33]

Surveys of the legal profession are quit common nowadays and provide insights into the economic realities faced by its members. One of the first such studies was conducted in 1934 providing a profile of the bar of New York on the basis of gender, race, years in practice, income, income variation on the basis of training and time in practice, practice organization (solo versus partnership), legal education, and type of clients.

Studies on legal education focused in two themes. One set of studies tried to determine what would be the best method of selecting law students for admission,[34] which has led to today's Law School Aptitude Test (LSAT). The other category of studies was focused on grading, especially differences in standards.[35]

How judges are being selected was and is a controversial issue. The first study in the United States created a ranking of the states as to the selection of judges based on a variety of characteristics for judges, such as education, marital status, number of children, age, religion, wealth, years of judicial service, organizational memberships, political experience, and a published rating of legal ability, which was benchmarked with a survey among law professors in each state rating the prestige of each Supreme Court.[36] A second study focused on the influence of the bar on the outcome of judicial elections in Chicago.[37] The researcher made an analysis of candidate factors (such as age, education, experience, etc.) that correlated with both the evaluation of candidates through a poll of members of the bar and the votes the candidates actual received.

---

30. Rodney L. Mott (1936), "Judicial Influence," *American Political Science Review* 30: 295-315.
31. Walton J. Wood (1916), "Necessity of Public Defender Established by Statistics," *Journal of the American Institute of Criminal Law & Criminology* 7: 230-244.
32. Reginald Heber Smith (1919), *Justice and the Poor*. New York: Carnegie Endowment for the Advancement of Teaching.
33. Charles E. Clark and Emma Corstvet (1938), "The Lawyer and the Public: An A.A.L.S. Survey," *Yale Law Journal* 47: 1272-1293.
34. See, for example, William L. Eagleton (1932), "Academic Preparation for Admission to Law School," *Illinois Law Review* 26: 607-644; and Frank Murray (1938), "Requirements for Admission to Law School," *Kentucky Law Journal* 26: 290-297.
35. See, for example, John Grant (1929), "The Single Standard in Grading," *Columbia Law Review* 29: 920-955.
36. Rodney L. Mott, Spencer D. Albright and Helen R. Semmerling (1933), "Judicial Personnel," *Annals of the American Academy of Political and Social Science* 167: 145-155.
37. Edward M. Martin (1936), *The Role of the Bar in Electing the Bench in Chicago*. Chicago: University of Chicago Press; and Edward M. Martin (1936), "The Selection of Judges in Chicago and the Role of the Local Bar Therein," *American Political Science Review* 30: 315-323.

Generally, the correlations were lower for the electorate than for the bar poll. Bar-endorsed candidates were more successful for Circuit and Superior Courts for the Municipal Court.[38]

§1.03 ELS AS AN ACADEMIC DISCIPLINE SUI GENERIS

ELS are by no means the successor of the legal realism movement that started in the first half of the 20th century or the disciplines of law and society and law and economics, because they all still exist. Actually, a significant portion of early ELS work was connected to the legal realism movement[39] and the origins of ELS predates the law and society and law and economics movements. It appears better to conceive ELS as a discipline not competing with the other disciplines but operating on a complementary track with them. ELS scholars use tools and methodology (for more detail, see section 1.04) generally applied in the social sciences, in particular statistical analysis. Traditionally, empirical legal scholars were trained in the social sciences, be it in sociology, economics, political science, finance and criminology, which led, notwithstanding their related scholarly interest, to a segmentation of empirical legal research as demonstrated by the isolation of law and economics from law and society.[40] According to Eisenberg, ELS has provided for a multidisciplinary center for law-related empirical scholarship "in law schools, in conferences, in a journal, and in other activities without sacrificing the distinct contribution of each discipline engaged in empirical scholarship."[41] Many empirical legal studies are now undertaken by lawyers with an additional graduate degree in another discipline.

Several factors can be distinguished in explaining the recent growth of empirical legal research. First, the maturity of legal research in general called for empirical testing of the theoretical and doctrinal assumptions and hypotheses. The increase in sheer volume of legal research and the bandwidth of issues considered by legal scholars naturally leads to more empirical testing possibilities. In that respect, the growth of the number of outlets for legal scholarship increases the variety of types of research, including the chances for empirical legal studies to reach publication. In direct addition, there is an increased demand for empirical research from the bench.

Proponents of legal empirical research, which is based on observation and experience, claim that traditional methods are enhanced by the employment of a solid base of up-to-date empirical information,[42] and that empirical research has the ability to help legal scholars determine whether a particular law or process is actually

---

38. See Kritzer, *supra* note 6, at 957.
39. Kritzer, *supra* note 6, at 926.
40. Victoria Nourse and Gregory Shaffer (2009), "Varieties of New Legal Realism: Can a New World Order Prompt a New Legal Theory?," *Cornell L. Rev.* 95: 61, 64. This isolation was also caused by the ideological foundation of the two movements, law and society being of liberal signature and law and economics having conservative roots.
41. Theodore Eisenberg, *supra* note 5, at 6.
42. Heise, *supra* note 4, at 824.

achieving its stated objectives.[43] After all, law is in essence the reflection of a society's moral values and the vehicle by which its economic, political and social objectives are to be delivered. A reality check makes sense to establish whether assumptions on deliverables are, indeed, correct and, regardless of good intentions, whether the law as instrument of policy lives up to expectations and delivers on promises. ELS researches the impact of law, compares results with objectives to measure success, identifies unintended consequences, thus delivering valuable information as a basis for policy-makers to instigate reform (when needed) and for lawyers, judges and others to improve their role performance in administering justice.

Legal scholars over the past 20 years have been encouraged to reflect on the role of empirical research, addressing important questions as to the scope of ELS, methodology of data gathering and analysis, which laws and legal processes are most suitable for research, and how the results of ELS can be elevated to the attention of the de facto owners of the political decision-making process.

The use of statistical analysis in individual cases has its limits in that the legal system requires a direct causation where statisticians and social scientists base their findings on statistical association without immediate concern about "whether the association corresponds with actual causation in a particular observation."[44]

The contribution of ELS will not be statistical analyses for the purpose of helping in deciding individual cases, but rather influencing policy questions and issues by providing evidence on how the legal system actually operates. Given the growth in online databases and their accessibility through the Internet, and the availability of inexpensive but sophisticated statistical software programs, an increase in ELS can be expected.

§1.04   RESEARCH METHODOLOGY

Empirical legal scholars have introduced and promote the application of social scientific analysis in legal academia. Mostly applied is a specific type of empirical research: a model-based approach coupled with a quantitative method. The empirical legal scholar offers a positive theory of a law or legal institution and then tests that theory using quantitative techniques developed in the social sciences. The evidence may be produced by controlled experiment or collected systematically from real world observation. In either event, quantitative or statistical analysis is a central component of the project.

Given the crucial role of judges in the economic, political, economic and social lives in a country, it comes as no surprise that a considerable percentage of contemporary empirical legal research is focused on the judicial decision-making process, i.e., what facts determine or explain judges' decisions. According to Heise over the last

---

43. N. William Hines, *Empirical Research: What Should We Study and How Should We Study It?* Association of American Law Schools, http://www.aals.org/am2006/theme.html (last visited February 7, 2011).
44. Theodore Eisenberg (2000). "Empirical Methods and the Law," *J. American Statistical Association* 95: 665, 666.

three decades studies use predominantly two research models, namely behavioralism and attitudinalism.[45] The first investigates the role of a judge's social background or personal attributes on judicial decisions.[46] However, empirical tests of behavioralism have led to mixed results. In other words, social background, gender, race, etc. seem hardly, or to a limited extent only, to explain judicial decisions.[47] Attitudinalism, the dominant research model according to Heise,[48] is a critical variation of behavioralism. Behavioralism presumes a direct influence on judicial decisions by a judge's socialeconomic background, where attitudinalism ascribe an indirect influence, i.e., someone's background is formative on his ideology which, in turn, would influence a judge's decision. Indeed, political affiliation has frequently identified as a determining factor — and thus a predictor — of judicial voting in ideologically divisive cases, but the difference overall between judges affiliated with the Democratic party and the Republican party respectively was relatively small.[49]

The limitations of both models have led to the development of more comprehensive models to explain and predict legal decision making. The legal model points to the moderating influence of legal doctrine and precedence as well as a judge's role orientation as judicial officer, which would lead to increased impartiality and objectivity among judges. The public choice theory suggests that in addition to the socio-economic background of judges, the legal context of the case and their actual reasoning should be taken into account, because judges as rational persons pursue instrumental and consumptive goals like everyone else.[50] Institutionalism focuses on how judges are influenced by the institutional characteristics within which they operate. Instead of assuming that and studying how judges influence the institutions in which they work, this approach takes the opposite angle and studies the "cultural" impact of the institution on the judge. Judges do not operate in isolation but function in intersecting relationships with other judges and therefore they exercise strategic behavior.[51]

Nevertheless, even the more developed research designs have their limitations. One limitation is the scope of the research. Many judicial decision-making studies are confined to published cases, which are by far in the minority. To use an example given by Heise,[52] only 4% of filed tort cases result in verdicts, and only some of those will be published. Another methodological limitation is found when scholars examine sample

---

45. Heise, *supra* note 4, at 833.
46. Frederick Schauer, Incentives, Reputation, and the Inglorious Determinants of Judicial Behavior, 68 *U. Cin. L. Rev.*: 2000: 615, 619.
47. Sue Davis *et al.*, Voting Behavior and Gender on the U.S. Court of Appeals, 77 *Judicature* 129, 130.
48. Heise, *supra* note 4, at 836.
49. See, for example, C.K. Rowland and Robert A. Carp (1996), "Politics and Judgment in Federal District Courts," University Press of Kansas: 24-57.
50. Richard A. Posner (1993), "What Do Judges Maximize? (The Same Thing Everybody Else Does)," *Sup. Ct. Econ. Rev.* 3(1): 1.
51. See, for example, Howard Gilman (1997), "The New Institutionalism, Part I: More and Less Than Strategy: Some Advantages to Interpretative Institutionalism in the Analysis of Judicial Politics," *Law & Cts.* 7 (1): 6; and Lee Epstein and Jack Knight (1997), "The New Institutionalism, Part II," *Law & Cts*, 7(2): 4.
52. Heise, *supra* note 4, at 844.

groups and series of unrelated cases. Studies covering a longer period of time seem to have an advantage over short-term research. However, the challenge here is to ensure that cases over a long period of time are sufficiently similar; otherwise, the results will not be comparable and generalizable. The frequency of a judge's vote for or against a defendant in criminal cases does not necessarily reflect a general attitude but may very well be the result of the unique set of facts in each particular case.[53] To improve comparability, researchers employ simulated cases. Simulated cases, however, miss authenticity in that experiments are conducted under conditions comparable to those in a laboratory and there exists no certainty that participants would behave exactly the same in a real case, especially since they are well aware that their decisions have no bearing on the resolution of an actual legal controversy. Moreover, many studies are focused on comparing outcomes, not on comparing legal reasoning.

A study is as good as the data on which it is based. Data need to be properly drawn and defined, also to allow for the research to be replicated by others. The selection of data is critical as well, where limitations in selection may negatively influence the value of the research. All serious scholars agree that empirical work needs to adhere to appropriately defined scholarly rules, standards and norms. Another risk is found in the potential bias of a researcher towards a certain outcome, in other words a researcher reads and interprets the results of his study with the objective to prove a certain theory and ignores elements of the data that potentially falsifies the theory. Most researchers are aware of that risk of course, and rigorous peer review should eliminate insufficiently founded conclusions.

Methodology is, indeed, the crucial element in empirical legal research and the discussion above indicates that there exist reasons to be cautious with the outcome and conclusions of ELS. There is reason for caution, not for skepticism, because researchers realize the limitations of and challenges posed by their chosen methodology, and are in constant quest to improve the design and reliability of their research.

Although the use of statistical tools is a characteristic of modern empirical legal research, it is correct to note that there are many other studies, which can justifiably be characterized as empirical in nature, but do not apply statistics. Many non-statistical economic studies are to a large extent also empirical and have a bearing for policy evaluation. Good examples in the field of taxation are incidence studies, determining to what degree the economic burden of a tax is, indeed, resting on the persons intended by the legislature to carry the tax.

A clear objective of ELS is to affect policy making, but its true influence is still relatively limited, partly because it is so difficult to conduct truly unassailable empirical research. Any doubt in respect of the accuracy of the research understandably reduces its credibility and thus its influence. Notwithstanding the structural and inherent limitations mentioned afore, ELS is bound to enhance and complement traditional legal scholarship, especially with the emergence of increasingly sophisticated statistical tools and more powerful computers.

---

53. Heise, *supra* note 4, at 845.

Chapter 1: Empirical Legal Studies and Taxation in the United States §1.05

§1.05 ELS STUDIES IN THE AREA OF TAXATION

The UCLA law school website[54] includes a bibliography of ELS articles, subdivided by category. It includes five tax categories: tax enforcement, tax policy, tax practice, tax reform and taxation. There are a total of 51 articles listed in all five categories but some are listed in more than one category. Thus, the total number of articles is less than 51. Brief summaries of each article are given in Table 1.1. They are arranged alphabetically by last name of first author.

Table 1.1   Summaries of ELS Tax Articles

Adda, Jerome and Francesca Cornaglia (2006). "Taxes, Cigarette Consumption, and Smoking Intensity." *Am. Econ. Rev.* 96: 1013.

> The authors used data from the National Health and Nutrition Examination Survey and cigarette excise tax data to evaluate the effect of taxes on cigarette and nicotine consumption using both the number of cigarettes consumed and cotinine levels. Regressions controlled for age, sex, race and several other demographic variables. The study found that a one percent increase in tax lead to a 0.47% increase in smoking intensity. Smokers adjust to the number of cigarettes they smoke and also to the amount of nicotine they extract. Heavier smokers were more prone to compensatory behavior. Since raising taxes causes people to smoke more intensively, the authors question the usefulness of tax increases.

Ahmed, Eliza and Valerie Braithwaite (2007). "Higher Education Loans and Tax Evasion: A Response to Perceived Unfairness." *Law & Pol'y* 29: 121.

> The Australian Higher Education Contribution Scheme (HECS) gives students the option of paying tuition up front or deferring payment until their income reaches some threshold level. The authors found that students who chose the deferral option were more likely to evade taxes by concealing income or overstating deductions. The purpose of the article was to explore the reasons why those who defer are more likely to evade taxes. Variables examined perceived unfairness of the HECS policy, satisfaction with the university course and the feeling of a moral obligation to pay taxes. Those who evaded were more likely to view the HECS policy as being unfair and were less satisfied with their university courses. They were also more dismissive of the tax authorities.

Barr, Michael S. and Jane K. Dokko (2008). "Third-Party Tax Administration: The Case of Low- and Moderate-Income Households." *J. Empirical Legal Stud.* 5: 963.

---

54. https://apps.law.ucla.edu/els/.

The authors used data from the Detroit Area Household Financial Services (DAHFS) study to examine the tax filing experiences of lower and middle income households. Tax filing behavior, attitudes about the withholding system, the use of tax refunds to save and consume and the mechanisms used to receive refunds were examined. A large percentage of taxpayers used paid preparers. About 38% of those who used a paid preparer took out refund anticipation loans (RAL). Banked and unbanked households were equally likely to receive tax refunds but banked households were 13% less likely to qualify for the earned income tax credit (EITC). Unbanked households were about 20% more likely to use a national chain to prepare their taxes. Unbanked households were more than twice as likely to take out a RAL, mostly because they would receive cash faster than waiting for their refund from the IRS. Nearly half of respondents said they took the RAL to pay the preparer. The authors conclude that efforts should be made to reduce tax complexity, since it is costing low and middle income taxpayers to pay high fees to have their taxes prepared. They also conclude that the tax system may encourage saving, since some taxpayers save their refunds.

Batchelder, Lily L., Fred T. Goldberg, Jr. and Peter R. Orszag (2006). "Efficiency and Tax Incentives: The Case for Refundable Tax Credits." *Stan. L. Rev.* 59: 23.

The authors argue that providing larger tax incentives to higher income groups is economically inefficient unless policy makers know that such households are more responsive to the incentive or unless taking advantage of the incentive leads to larger social benefits. They propose a uniform refundable tax credit as the default method for providing incentives rather than the current system, which favors those in higher tax brackets.

Battaglini, Marco and Stephen Coate. "Inefficiency in Legislative Policymaking: A Dynamic Analysis." *Am. Econ. Rev.* 97: 118.

The authors test the hypothesis that legislators will act inefficiently by overspending on pork barrel projects and under spending on public goods. The study provides support for the conventional wisdom that legislators from geographic specific districts will produce a government that is too large and a level of public goods that is too low. They qualify this general tendency by concluding that when the capacity to tax is low, politicians tend to be more efficient by providing more public goods than would otherwise be the case. They will be less likely to push for pork barrel projects when the result would be to distort the economy.

Bento, Antonio, Lawrence H. Goulder, Mark R. Jacobsen and Roger H. Von Haefen (2009). "Distributional and Efficiency Impacts of Increased US Gasoline Taxes." *Am. Econ. Rev.* 99: 667.

> The authors examine the option of increasing gasoline taxes using an econometrically based multimarket simulation model to evaluate the policy's efficiency and distributional effects. An investigation was made of increased US gasoline taxes on fuel consumption and the shift to higher mileage automobiles and explored how the costs were distributed across households that differ by income, race, region and other demographics. The study found that each one cent per gallon increase in the gasoline tax reduced consumption by 0.2%. The impact of a 25 cent per gallon tax increase was about USD 30 per year for the average household. The distributional impact depended on how the additional revenues from the tax increase were recycled.

Berry, Christopher (2008). "Piling On: Multilevel Government and the Fiscal Common-Pool." *Am. J. Pol. Sci.* 52: 802

> The author examined the common pool problems that arise when overlapping governments share the authority to tax and provide services in a common geographic area. Using regression analysis the author found that the effect of jurisdictional overlap is positive and highly significant. Increasing the number of overlapping jurisdictions results in "overfishing" from the shared tax base and increases the public sector.

Bertrand, Marianne and Adair Morse (2009). "What Do High-Interest Borrowers Do with Their Tax Rebate?" *Am. Econ. Rev.* 99: 418

> The authors explore the issue of what high interest borrowers do with their tax rebates. They examine payday loan customers, for whom the cost of marginal debt can be extremely high, more than 400% APR. The study found that borrowing declined significantly in the pay periods following receipt of a tax rebate. There was a tendency to use the tax rebate for vacations, apparel, electronics, gifts or entertainment rather than for retiring debt or saving. Those who use the payday loan option less often tend to be more likely to use the tax rebate to retire debt. High frequency users did not retire debt.

Braithwaite, Valerie, Kristina Murphy and Monika Reinhart (2007). "Taxation Threat, Motivational Postures & Responsive Regulation." *Law & Pol'y* 29: 137

> Motivational posturing theory holds that those being regulated place themselves at a distance from those who are doing the regulating. This study collected data from more than 3,000 Australians, many of whom were having difficulties with the tax authority. They examined three motivational postures: thinking morally, feeling oppressed and taking control. Resistance-cooperation was measured by aggregating responses to 18 items using a five-point Likert Scale. Dismissiveness examined responses to 10 items. Data were analyzed using factor analysis and varimax rotation. The authors take the position that the most effective regulatory outcome is achieved when the regulatory process is able to dampen the "taking control" and "feeling oppressed" attitudes and strengthen the "thinking morally" attitude.

Cagetti, Marco and Mariacristina De Nardi (2009). "Estate Taxation, Entrepreneurship, and Wealth." *Am. Econ. Rev.* 99: 85

The authors examine the effects on general welfare of abolishing the estate tax under three different scenarios: no tax replacement but cuts in wasteful spending; replacing the estate tax revenue losses with a consumption tax or a higher income tax. The authors conclude that the estate tax has little effect on the saving and investment decisions of small businesses but does distort the decisions of larger businesses, thus reducing aggregate output and savings. If the estate tax were replaced with another form of taxation, those at the top of the income scale would experience a large welfare gain while those at lower levels would experience a welfare loss.

Clarkson, Gavin (2007). "Tribal Bonds: Statutory Shackles and Regulatory Restraints on Tribal Economic Development." *N.C. L. Rev.* 85: 1009.

Tax law restricts the ability of American Indian tribes from raising revenue using tax exempt bonds. These restrictions have severely limited their ability to access the capital markets. These restrictions harm the poorer tribes the most and often make it impossible to address infrastructure deficiencies. Tribal governments are also targeted disproportionately in IRS audits of tax-exempt municipal offerings (30%, compared to less than 1% for non-tribal offerings), and 100% of tribal conduit issuances are being challenged by the IRS. The author argues that tribal governments should have the same bonding authority as their state and local counterparts and that expanding tribal bonding authority would increase federal revenues.

Cooper, Graeme S. and Michael Wenzel (2009). "Testing Alternative Legal Paradigms: An Experiment in Designing Tax Legislation." *Law & Soc. Inquiry* 34: 61.

The authors conducted empirical research to test the claim that citizens could be more certain of their legal obligations by changing the legal paradigms used to express their rights and obligations. Tests were made of a number of hypotheses that involved different formulations of the claim being made. The study found that the alternative paradigm was inferior to current practice. Some reasons were offered to explain the results. There was also a discussion of the significance of the study for other areas of legal research.

Edlund, Lena and Wojciech Kopczuk (2009). "Women, Wealth, and Mobility." *Am. Econ. Rev.* 99: 146.

This study examined estate tax data and found that the share of very wealthy women in the United States peaked at nearly one-half in the late 1960s, then declined to one-third. The authors claim that this decline reflects changes in the importance of dynastic wealth. They conclude that wealth mobility declined until the 1970s, then rose. One of the underlying assumptions was that men tend to earn the wealth whereas both men and women inherit it.

Gehlbach, Scott (2006). "The Consequences of Collective Action: An Incomplete-Contracts Approach." *Am. J. Pol. Sci.* 50: 802.

The author surveyed firms in 25 post-communist countries using data from the Business Environment and Enterprise Performance Survey carried out by the World Bank and the European Bank for Reconstruction and Development to determine the relationship between 16 variables and the probability of hiding revenue from the government. The author found that there is sometimes a significant correlation between variables and tax compliance and that attitudes towards tax compliance in Eastern Europe and the Baltics sometimes differs significantly from attitudes in the CIS countries (other than the Baltics).

Glass, Anthony. "Government Expenditure on Public Order and Safety, Economic Growth and Private Investment: Empirical Evidence from the United States." *Int'l Rev. L. & Econ.* 29: 29.

The author uses data from the United States from 1959 to 2003 to examine the relationship between government expenditure on public order and safety, economic growth and private investment. He tests for Granger causality between output, investment and each category of spending (police force, fire service, law courts and prison service). Although there is a lot of evidence on unidirectional causality, the only case of bi-directional causality involved changes in spending on the police force and changes in expenditures on the law courts.

Hickman, Kristin E. "Coloring outside the Lines: Examining Treasury's (Lack of) Compliance with Administrative Procedure Act Rulemaking Requirements." *Notre Dame L. Rev.* 82: 1727.

This article examined 232 regulatory projects interpreting the Internal Revenue Code to determine the extent of the Treasury Department's compliance with the Administrative Procedures Act's requirement of allowing a period for public comment on proposed regulatory changes and found that the Treasury often does not comply with the APA requirements, thus leaving many regulations open to legal challenge.

Hopkins, Daniel J (2009). "The Diversity Discount: When Increasing Ethnic and Racial Diversity Prevents Tax Increases." *J. Pol.* 71: 160

The author studied the impact of racial and ethnic demographics on property tax votes in Massachusetts and Texas using time-series cross-sectional data. He found that diversity reduces a locality's willingness to raise taxes only when the locality is undergoing sudden demographic changes. Where a locality has stable demographics, there does not seem to be a barrier to raising taxes.

House, Christopher L. and Matthew D. Shapiro (2006). "Phased-In Tax Cuts and Economic Activity." *Am. Econ. Rev.* 96: 1835.

This study considers the macroeconomic implications of the timing of tax cuts. Using a dynamic general equilibrium model, the authors conclude that the timing of tax cuts have a substantial effect on output, labor and investment. The slow recovery from the 2001 recession in the United States was found to be partially attributable to the phase-in of tax cuts (perhaps immediate tax cuts would have been better). Phased-in tax cuts provide incentives to delay increases in economic activity.

Johnson, David S., Jonathan A. Parker and Nicholas S. Souleles. "Household Expenditures and the Income Tax Rebates of 2001." *Am. Econ. Rev.* 96: 1589.

> The authors estimate the change in consumption expenditures caused by the 2001 federal income tax rebates and test the permanent income hypothesis. They found that households spent 20 to 40 percent of their rebates on nondurable goods within three months of the receipt of their rebates and roughly two-thirds of their rebates within six months of receipt. Responses were larger for low income households and households with low liquid wealth.

Kasara, Kimuli. "Tax Me If You Can: Ethnic Geography, Democracy, and the Taxation of Agriculture in Africa." *Am. Pol. Sci. Rev.* 101: 159.

> This study casts doubt on the theory that African leaders enact policies that benefit members of their ethnic group at the expense of other groups. Cash crop farmers who are of the same ethnic group as the leader were found to pay higher taxes than do members of other ethnic groups. One reason given for this preferred treatment is to reduce the possibility that members of rival ethnic groups will support opposition leaders. The study also found that those who live in democratic regimes pay lower taxes than those who live in nondemocratic regimes.

Khwaja, Ahmed, Frank Sloan and Yang Wang (2009). "Do Smokers Value Their Health and Longevity Less?" *J. L. & Econ.* 52: 171.

> The authors examined the internal costs of chronic obstructive pulmonary disease and found that smokers face a lower full cost of smoking because of different preferences than nonsmokers. The implications of the study suggest that the case for intervention in the form of regulation or taxation aimed at helping smokers to quit may be weak.

Mann, Ronald J. and Katherine Porter. "Saving up for Bankruptcy." *Geo. L.J.* 98: 289.

> This study attempts to determine why some people in financial distress file for bankruptcy while others do not, even though they could benefit by filing. The study found that the probability of filing increases if the debtor is subject to aggressive collection techniques. People time the filing of their bankruptcy petition based on when they are able to save enough money to pay their attorney and filing fees, which is often shortly after receiving their income tax refund check.

Sanchirico, Chris William (2008). "Progressivity and Potential Income: Measuring the Effect of Changing Work Patterns on Income Tax Progressivity." *Colum. L. Rev.* 108: 1551.

> The author takes nearly 60 pages to prove that, as marginal tax rates increase, people tend to work less at taxable activities. Although marginal tax rates increased in the 1990s in the United States, progressivity actually declined, as individuals tended to increase the amount of time engaged in nonmarket activities, such as staying home to care for their children rather than outsourcing it.

Schochet, Peter Z., John Burghardt and Sheena McConnell (2008). "Does Job Corps Work? Impact Findings from the National Job Corps Study." *Am. Econ. Rev.* 98: 1864.

Using tax and other data collected for up to nine years, the authors found that the Job Corps program, which is aimed at disadvantaged youths, tended to enhance educational attainment, reduce crime rates and increase earnings for several years after leaving the program. However, the earnings gains were not sustained except for older participants.

Shapiro, Matthew and Joel Slemrod (2009). "Did the Tax Rebates Stimulate Spending?" *Am. Econ. Rev.* 99: 374.

The authors examined the effect of the Economic Stimulus Act of 2008 on consumer spending using data gathered from a University of Michigan consumer survey. Nearly half of the respondents said they would use the rebate mostly to pay off debt; slightly more than 30% said they would save most of it; nearly 20% said they would spend most of it. Those over age 65 were more likely to spend than people in younger age groups. The correlation of income to spending was weak, but individuals in the lowest income group tended to spend slightly less of their rebate than did people in most of the higher income groups, a finding that casts aspersions on the theory that the marginal propensity to consume is higher for low income groups.

Slemrod, Joel. Why Is Elvis on Burkina Faso Postage Stamps? Cross-Country Evidence on the Commercialization of State Sovereignty." *J. Empirical Legal Stud.* 5: 683.

The author conducted an empirical investigation to determine the circumstances under which a nation will engage in the commercialization of state sovereignty, such as the selling of stamps that have value to collectors, becoming a tax haven and engaging in money laundering. The study found that countries are more likely to do so when alternative ways of raising funds are difficult. Stamp pandering and tax havens were more popular in poorer countries and in smaller countries.

Waller, Vivienne (2007). "The Challenge of Institutional Integrity in Responsive Regulation: Field Inspections by the Australian Taxation Office." *Law & Pol'y* 29: 67.

This study demonstrates the unintended consequences that can occur when there is a lack of institutional integrity at the design level and the level of field officer practices. The study involved interactions between the Australian Tax Office and used car dealers. Information was gathered by observation and interviews. Unannounced walk-ins by the tax authorities often tended to be counterproductive.

Weber, Anke (2009). "An Empirical Analysis of the 2000 Corporate Tax Reform in Germany: Effects on Ownership and Control in Listed Companies." *Int'l Rev. L. & Econ.* 29: 57

> The author conducted an empirical analysis of the 2000 corporate tax reform in Germany and its effects on the ownership and control of listed companies. The study found a fall in ownership concentration and a decrease in the power of institutional owners, including banks. The author suggests the possibility that German corporate governance may no longer be bank based. However, the change in the tax law did not result in a revolution in German corporate governance. Concentration of ownership is still high compared to the Anglo-American countries. The author did not observe an active market for corporate control.

Wenzel, Michael (2007). "The Multiplicity of Taxpayer Identities and Their Implications for Tax Ethics." *Law & Pol'y* 29: 31.

> The study presents the results of an Australian taxpayer survey of tax ethics and included demographic factors such as gender, age, education level and income level. The study found that those who most strongly identified themselves as Australians (members of a national group) had more positive views towards tax compliance. Those who most strongly identified themselves with some subgroup, such as a member of a certain income level, tended to be less tax compliant. Feelings of shame and guilt are correlated with tax compliance.

The UCLA Law School website is incomplete. It does not list all the empirical studies that have been made of tax topics. Table 1.2 lists another group of studies, all of which deal with the ethics of tax evasion.[55] The studies in Table 1.2 all used a similar survey instrument. It consisted of a series of 15-18 statements beginning with the phrase, "Tax evasion is ethical if ..." and contained a seven-point Likert scale ranging from "strongly agree" to "strongly disagree" to indicate the extent of agreement with each statement. A ranking of arguments was then made to indicate the relative strength or weakness of the various arguments. Demographic analyses were sometimes done to determine whether responses differed by demographic variable.

In many cases, the strongest arguments had to do with human rights abuses. The three human rights statements were "Tax evasion is ethical if ..." (1) I were a Jew living in Nazi Germany; (2) the government imprisons people for their political opinions; and (3) the government discriminates against me because of my religion, race or ethnic background. The three human rights statements were omitted from the Chinese surveys.

---

55. Summaries of some of these studies may be found at http://ssrn.com/author=2139 and also in Robert W. McGee (ed.) (2012), *The Ethics of Tax Evasion in Theory and Practice*. New York: Springer.

Table 1.2 Summary of Tax Evasion Studies Using 15-18 Statement Survey Instrument

Citation	Summary
McGee, Robert W. and Tatyana B. Maranjyan (2012). "Attitudes toward Tax Evasion: An Empirical Study of Florida Accounting Practitioners." In Robert W. McGee (ed.), The Ethics of Tax Evasion in Theory and Practice. New York: Springer.	Accounting Practitioners in Florida: Strongest arguments to justify tax evasion in cases of human rights abuses. Women were slightly more opposed to tax evasion. Non-Hispanic whites were significantly more opposed to tax evasion than were Hispanic whites.
McGee, Robert W. and Marcelo J. Rossi (2008). "A Survey of Argentina on the Ethics of Tax Evasion." In Robert W. McGee (ed.), *Taxation and Public Finance in Transition and Developing Economies* (pp. 239-261). New York: Springer.	**Argentina Business, Economics & Law Students**: The strongest arguments to justify tax evasion were inability to pay, human rights abuses and corruption. Students and faculty were equally opposed to tax evasion. Men and women were equally opposed to tax evasion. Business and economics students were more opposed to tax evasion than were law students.
McGee, Robert W. and Tatyana B. Maranjyan (2008). "Opinions on Tax Evasion in Armenia." In Robert W. McGee (ed.), *Taxation and Public Finance in Transition and Developing Economies* (pp. 277-307). New York: Springer.	**Armenia Business, Economics & Theology Students**: The strongest arguments to justify tax evasion were in cases where the system was perceived to be unfair or where there was corruption or waste. Business students were more strongly opposed to tax evasion than were theology students.
McGee, Robert W. and Sanjoy Bose (2009). "The Ethics of Tax Evasion: A Survey of Australian Opinion." In Robert W. McGee (ed.), *Readings in Business Ethics* (pp. 143-166). Hyderabad, India: ICFAI University Press.	**Australia Business, Philosophy and Seminary Students and Faculty**: The strongest arguments to justify tax evasion were in cases where tax rates were too high, where the system was perceived to be unfair or where tax funds were wasted. Females were somewhat more opposed to tax evasion. Undergraduate students were least opposed to tax evasion. Faculty was most opposed to tax evasion. Business and economics students were least opposed to tax evasion; seminary students were most opposed. Muslims were least opposed to tax evasion; Catholics were most opposed. Asians were least opposed to tax evasion. Anglos were most opposed.

Citation	Summary
McGee, Robert W., Meliha Basic and Michael Tyler (2009). "The Ethics of Tax Evasion: A Survey of Bosnian Opinion." *Journal of Balkan and Near Eastern Studies* 11(2): 197-207.	**Bosnia & Herzegovina Advanced Undergraduate Business & Economics Students**: The strongest arguments to justify tax evasion were in cases where there was government corruption or where the government engaged in human rights abuses.
McGee, Robert W., Meliha Basic and Michael Tyler (2008). "The Ethics of Tax Evasion: A Comparative Study of Bosnian and Romanian Opinion." In Robert W. McGee (ed.), *Taxation and Public Finance in Transition and Developing Economies* (pp. 167-183). New York: Springer.	**Bosnian & Romanian Business Students**: The strongest arguments to justify tax evasion were in cases where there was government corruption or where the government engaged in human rights abuses in the case of Bosnian students and in cases of unfairness, discrimination or high tax rates in the case of Romanian students. Bosnians were generally more opposed to tax evasion.
McGee, Robert W. and Zhiwen Guo (2007). "A Survey of Law, Business and Philosophy Students in China on the Ethics of Tax Evasion." *Society and Business Review* 2(3): 299-315.	**Chinese Graduate and Advanced Undergraduate Business & Economics, Law and Philosophy Students in Central China**: The strongest arguments to justify tax evasion were in cases where there is corruption or waste (the human rights statements were omitted from all Chinese surveys). Women were significantly more opposed to tax evasion. Business & economics students were least opposed to tax evasion; law and philosophy students were equally opposed to tax evasion.
McGee, Robert W. and Yuhua An (2008). "A Survey of Chinese Business and Economics Students on the Ethics of Tax Evasion." In Robert W. McGee (ed.), *Taxation and Public Finance in Transition and Developing Economies* (pp. 409-421). New York: Springer.	**Graduate and Advanced Undergraduate Business & Economics Students in Beijing**: The strongest arguments to justify tax evasion were in cases where there is corruption or waste (the human rights statements were omitted from all Chinese surveys). Men and women were equally opposed to tax evasion.

Citation	Summary
McGee, Robert W. and Carlos Noronha (2008). "The Ethics of Tax Evasion: A Comparative Study of Guangzhou (Southern China) and Macau Opinions." *Euro Asia Journal of Management* 18(2): 133-152.	**Southern China & Macau Social Science, Business & Economics Students**: The strongest arguments to justify tax evasion were in cases where the system is perceived as unfair or where there is corruption or waste (the human rights statements were omitted from all Chinese surveys). Mainland Chinese and Macau students were equally opposed to tax evasion. Men and women were equally opposed to tax evasion.
McGee, Robert W., Silvia López Paláu and Gustavo A. Yepes Lopez (2009). "The Ethics of Tax Evasion: An Empirical Study of Colombian Opinion." In Robert W. McGee (ed.), *Readings in Business Ethics* (pp. 167-184). Hyderabad, India: ICFAI University Press.	**Colombia Business Students**: The strongest arguments to justify tax evasion were in cases where there is corruption, waste or human rights abuses. Women were more opposed to tax evasion than were men.
McGee, Robert W., Silvia López-Paláu and Fabiola Jarrín Jaramillo (2007). "The Ethics of Tax Evasion: An Empirical Study of Ecuador." American Society of Business and Behavioral Sciences 14th Annual Meeting, Las Vegas, February 22-25, 2007. Published in the *Proceedings of the American Society of Business and Behavioral Sciences* 14(1): 1186-1198.	**Ecuador Business Students**: The strongest arguments to justify tax evasion were in cases where there is corruption, waste or unfairness. Women were slightly more opposed to tax evasion.
McGee, Robert W., Jaan Alver and Lehte Alver (2008). "The Ethics of Tax Evasion: A Survey of Estonian Opinion." In Robert W. McGee (ed.), *Taxation and Public Finance in Transition and Developing Economies* (pp. 461-480). New York: Springer.	**Estonia Graduate and Undergraduate Business Students, Faculty & Practitioners**: The strongest arguments to justify tax evasion were in cases where there is corruption, human rights abuses or the perception of unfairness. Women were significantly more opposed to tax evasion. Overall, undergraduate students were least opposed to tax evasion; faculty and practitioners were most opposed. People under age 25 were significantly less opposed to tax evasion than were people in the 25-40 age group. Accounting students and business & economics students were equally opposed to tax evasion.

Citation	Summary
McGee, Robert W., Jaan Alver and Lehte Alver (2012). "Tax Evasion Opinion in Estonia." In Robert W. McGee (ed.), *The Ethics of Tax Evasion in Theory and Practice*. New York: Springer.	**Estonia Accounting and Other Business Students and Practitioners**: The strongest arguments to justify tax evasion were in cases where there is corruption, human rights abuses or the perception of unfairness. Women were significantly more opposed to tax evasion. People in the oldest group (40+) were significantly more opposed to tax evasion than were people in the youngest group (under 25). Accounting practitioners were often significantly more opposed to tax evasion than either undergraduate or graduate students. Accounting majors were somewhat more opposed to tax evasion than were other business & economics majors.
McGee, Robert W. and Bouchra M'Zali (2009). "The Ethics of Tax Evasion: An Empirical Study of French EMBA Students." In Robert W. McGee, *Readings in Business Ethics* (pp. 185-199). Hyderabad, India: ICFAI University Press. An abbreviated version was published in Adams, Marjorie G. and Abbass Alkhafaji, (eds.), *Business Research Yearbook: Global Business Perspectives*, Volume XIV, No. 1 (Beltsville: International Graphics, MD., 2007), 27-33.	**France Executive MBA Students**: The three strongest arguments to justify tax evasion all had to do with human rights abuses. Men and women were equally opposed to tax evasion.
McGee, Robert W., Inge Nickerson and Werner Fees (2009). "When Is Tax Evasion Ethically Justifiable? A Survey of German Opinion." In Robert W. McGee (ed.), *Readings in Accounting Ethics* (pp. 365-389). Hyderabad, India: ICFAI University Press.	**German Graduate and Upper Division Undergraduate Business Students**: The strongest arguments to justify tax evasion were in cases of human rights abuses and corruption.

Chapter 1: Empirical Legal Studies and Taxation in the United States §1.05

Citation	Summary
McGee, Robert W., Inge Nickerson and Werner Fees (2006). "German and American Opinion on the Ethics of Tax Evasion." *Proceedings of the Academy of Legal, Ethical and Regulatory Issues* (Reno) 10(2): 31-34.	**German and US Business Students**: Tax evasion was most justified in cases of human rights abuses. American students were significantly more opposed to tax evasion than German students. American women were significantly more opposed to tax evasion than were American men.
McGee, Robert W. and Christopher Lingle (2008). "The Ethics of Tax Evasion: A Survey of Guatemalan Opinion." In Robert W. McGee, (ed.), *Taxation and Public Finance in Transition and Developing Economies* (pp. 481-495). New York: Springer.	**Guatemala Business, Economics and Law Students**: Tax evasion was more justified in cases where there was corruption, waste or human rights abuses. Women were more opposed to tax evasion than were men. Business students were more opposed to tax evasion than were law students.
McGee, Robert W. and Bouchra M'Zali (2012). Opinions on Tax Evasion in Haiti. In Robert W. McGee (ed.), *The Ethics of Tax Evasion in Theory and Practice*. New York: Springer.	**Haiti Business and Economics Students**: Tax evasion was more justified in cases where tax rates were too high, the system is perceived as unfair, there is waste or human rights abuses. Women were somewhat more opposed to tax evasion than were men.
McGee, Robert W., Arsen M. Djatej and Robert Zeke Sarikas. "The Ethics of Tax Evasion: A Survey of Hispanic Opinion." *Accounting & Taxation*, forthcoming.	**Hispanic Business Students in South Texas**: Tax evasion was more justified in cases where the government engages in human rights abuses or where the system is perceived as unfair. Attitude towards tax evasion did not differ by gender or age. Accounting students were significantly more opposed to tax evasion than were business and economics majors.
McGee, Robert W., Simon S.M. Ho and Annie Y.S. Li (2008). "A Comparative Study on Perceived Ethics of Tax Evasion: Hong Kong vs. the United States." *Journal of Business Ethics* 77(2): 147-158.	**Hong Kong & the US Business Students**: Tax evasion was more justified for Hong Kong students in cases where there was corruption or waste or where the proceeds go to support an unjust war. The strongest arguments to justify tax evasion for the US students were in cases of corruption, perceived unfairness of the tax system or inability to pay. Overall, the differences in mean scores between Hong Kong and US students were not significant, although there were significant differences for some of the statements.

Citation	Summary
McGee, Robert W. and Yiu Yu Butt (2008). "An Empirical Study of Tax Evasion Ethics in Hong Kong. Proceedings of the International Academy of Business and Public Administration Disciplines (IABPAD)." Dallas, April 24-27: 72-83.	**Hong Kong Business Students**: Tax evasion was more justified in cases where the system was perceived as unfair, where there was corruption or waste. Men and women were equally opposed to tax evasion.
McGee, Robert W. and Beena George (2008). "Tax Evasion and Ethics: A Survey of Indian Opinion." *Journal of Accounting, Ethics & Public Policy* 9(3), 301-332.	**India Graduate Business Students in Kerala**: Tax evasion was more justified in cases where the tax system was perceived as unfair or where there was waste or corruption. Men and women were equally opposed to tax evasion. Hindus were slightly more opposed to tax evasion than were Christians.
McGee, Robert W. and Ravi Kumar Jain (2012). "The Ethics of Tax Evasion: A Study of Indian Opinion." In Robert W. McGee (ed.), *The Ethics of Tax Evasion in Theory and Practice*. New York: Springer.	**India Accounting, Business and Engineering Students (mostly graduate) and Faculty**: Tax evasion was more justified in cases where there was waste or corruption or where the system was perceived as unfair. Men were slightly more opposed to tax evasion than were women. Faculty was significantly more opposed to tax evasion than were graduate students. Accounting students were slightly more opposed to tax evasion than were business and economics students.
McGee, Robert W. (2006). "A Survey of International Business Academics on the Ethics of Tax Evasion." *Journal of Accounting, Ethics & Public Policy* 6(3): 301-352.	**International Business Academics**: Tax evasion was more justified in cases where the government engaged in human rights abuses or there was corruption. Women were more opposed to tax evasion in all 18 cases.
McGee, Robert W. and Mahdi Nazemi Ardakani (2009). "The Ethics of Tax Evasion: A Case Study of Opinion in Iran." Florida International University Working Paper.	**Iran Master's Degree Accounting Students**: Tax evasion was more justified in cases where the system is perceived as being unfair or where there is corruption. Men and women were equally opposed to tax evasion.

Chapter 1: Empirical Legal Studies and Taxation in the United States §1.05

Citation	Summary
McGee, Robert W. and Gordon M. Cohn (2008). "Jewish Perspectives on the Ethics of Tax Evasion." *Journal of Legal, Ethical and Regulatory Issues* 11(2): 1-32.	**Orthodox Jewish Undergraduate Students in New York**: Tax evasion was more justified in cases where the government engaged in human rights abuses or where the government was corrupt. Women were significantly more opposed to tax evasion than were men.
McGee, Robert W. and Galina G. Preobragenskaya (2008). "A Study of Tax Evasion Ethics in Kazakhstan." In Robert W. McGee (ed.), *Taxation and Public Finance in Transition and Developing Economies* (pp. 497-510). New York: Springer.	**Kazakhstan Accounting and Business Students**: Tax evasion was more justified in cases where the government discriminates, where there is corruption or where the tax system is perceived as unfair. Men and women were equally opposed to tax evasion. Accounting and other business students were equally opposed to tax evasion.
McGee, Robert W., Carlos Noronha and Michael Tyler (2007). "The Ethics of Tax Evasion: A Survey of Macau Opinion." *Euro Asia Journal of Management* 17(2): 123-150. Reprinted in Robert W. McGee (ed.), *Readings in Accounting Ethics* (pp. 283-313). Hyderabad, India: ICFAI University Press, 2009.	**Macau Graduate and Undergraduate Business Students**: Tax evasion was more justified in cases where there is corruption, where the tax proceeds go to support an unjust war or where the system is perceived as being unfair (the three human rights issues were not included in this survey). Men and women were equally opposed to tax evasion.
McGee, Robert W. and Bouchra M'Zali (2008). "Attitudes toward Tax Evasion in Mali." In Robert W. McGee (ed.), *Taxation and Public Finance in Transition and Developing Economies* (pp. 511-517). New York: Springer.	**Mali Executive MBA Students**: Tax evasion was more justified in cases where the government wastes money, where the government engages in human rights abuses or where the tax system is perceived as unfair. Support for tax evasion was widespread.

Citation	Summary
McGee, Robert W., Yanira Petrides and Adriana M. Ross (2012). "Ethics and Tax Evasion: A Survey of Mexican Opinion." In Robert W. McGee (ed.), *The Ethics of Tax Evasion in Theory and Practice.* New York: Springer.	**Mexico Business and Engineering Students, Faculty and Nonstudents**: Tax evasion was more justified in cases where there is corruption or waste or where the tax system is perceived as unfair. Women were slightly more opposed to tax evasion than were men. The under 25 group was significantly less opposed to tax evasion than the other two groups. The 25-40 group was most opposed to tax evasion. The Over 40 group was between the other two groups. Engineering students were most opposed to tax evasion. Other business/economics students were least opposed to tax evasion. Accounting students were in between the other two groups. Nonstudents (workers) were more opposed to tax evasion than were any other group. Undergraduate students were least opposed to tax evasion. Faculty was more opposed to tax evasion than either graduate or undergraduate students.
McGee, Robert W. and Sheldon R. Smith (2007). "Ethics, Tax Evasion and Religion: A Survey of Opinion of Members of the Church of Jesus Christ of Latter-Day Saints." Western Decision Sciences Institute, Thirty-Sixth Annual Meeting, Denver, April 3-7. Published in the Proceedings. Reprinted at http://ssrn.com/abstract=934652.	**Mormon Business, Legal Studies and Technology Students**: Tax evasion was more justified in cases where the government engages in human rights abuses, where there is corruption or where the tax system is perceived as being unfair. Women were significantly more opposed to tax evasion than were men. Mormons were significantly more opposed to tax evasion than were non-Mormons in all 18 cases.

Chapter 1: Empirical Legal Studies and Taxation in the United States §1.05

Citation	Summary
Gupta, Ranjana and Robert W. McGee (2010). "A Comparative Study of New Zealanders' Opinion on the Ethics of Tax Evasion: Students v. Accountants." *New Zealand Journal of Taxation Law and Policy* 16(1): 47-84.	**New Zealand Graduate and Undergraduate Accounting, Business & Economics and Law Students and Accounting Practitioners**: Tax evasion was more justified in cases where there was corruption or waste or where the tax system was perceived as being unfair. Women were more opposed to tax evasion than were men. Older people were more opposed to tax evasion than were younger people. Graduate students were more opposed to tax evasion than were undergraduate students. Law students were somewhat less opposed to tax evasion than were the accounting and business/economics students. Accounting practitioners were significantly more opposed to tax evasion than were any other group. The European ethnic group was significantly more opposed to tax evasion than were the other two groups. The Asian and Pasifika groups were equally opposed to tax evasion. Catholics were most opposed to tax evasion; Buddhists were least opposed.
McGee, Robert W. (2012). "Attitudes on the Ethics of Tax Evasion: A Survey of Philosophy Professors." In Robert W. McGee (ed.), *The Ethics of Tax Evasion in Theory and Practice*. New York: Springer.	**Philosophy Professors**: Tax evasion was more justified in cases where the government engages in human rights abuses or where there is corruption or where the tax system is perceived as unfair. Women were significantly more opposed to tax evasion than were men.
McGee, Robert W. and Arkadiusz Bernal (2006). "The Ethics of Tax Evasion: A Survey of Business Students in Poland." In Mina Baliamoune-Lutz, Alojzy Z. Nowak and Jeff Steagall (eds.), *Global Economy — How It Works* (pp. 155-174). Warsaw: University of Warsaw & Jacksonville: University of North Florida.	**Poland Economics Students**: Tax evasion was more justified in cases where the government is corrupt or wasteful or where the government engages in human rights abuses. Men and women were equally opposed to tax evasion.

Citation	Summary
McGee, Robert W. and Silvia López Paláu (2007). "The Ethics of Tax Evasion: Two Empirical Studies of Puerto Rican Opinion." *Journal of Applied Business and Economics* 7(3): 27-47 (2007). Reprinted in Robert W. McGee (ed.) (2009). *Readings in Accounting Ethics* (pp. 314-342). Hyderabad, India: ICFAI University Press.	**Puerto Rico Accounting and Law Students**: Tax evasion was more justified in cases where there is corruption, waste or unfairness in the tax system. Women were more opposed to tax evasion than were men. Accounting students were more opposed to tax evasion in 9 of 18 cases; law students were more opposed in 9 of 18 cases.
McGee, Robert W. (2006). "The Ethics of Tax Evasion: A Survey of Romanian Business Students and Faculty." *The ICFAI Journal of Public Finance* 4(2): 38-68. Reprinted in Robert W. McGee and Galina G. Preobragenskaya (2006). *Accounting and Financial System Reform in Eastern Europe and Asia* (pp. 299-334). New York: Springer.	**Romania Graduate and Upper Division Undergraduate Business Students**: Tax evasion was more justified in cases where the tax system is perceived as being unfair, where the government discriminates or where the taxpayer is unable to pay. Men were slightly more opposed to tax evasion than were women.
Nickerson, Inge, Larry P. Pleshko and Robert W. McGee (2009). "Presenting the Dimensionality of an Ethics Scale Pertaining to Tax Evasion." *Journal of Legal, Ethical and Regulatory Issues*, 12(1): 1-14.	**Six Countries — US, Argentina, Guatemala, Poland, Romania, UK**: Tax evasion was more justified in cases where the government is corrupt or discriminates or where the tax system is perceived as being unfair. The United States sample was most strongly opposed to tax evasion, followed by Argentina, Guatemala, Poland, Romania and the United Kingdom.
McGee, Robert W. and Radoslav Tusan (2008). "The Ethics of Tax Evasion: A Survey of Slovak Opinion." In Robert W. McGee (ed.), *Taxation and Public Finance in Transition and Developing Economies* (pp. 575-601). New York: Springer.	**Slovakia Business & Economics, Philosophy and Theology Students**: Tax evasion was more justified in cases where the government engages in human rights abuses or where the government is corrupt. Men were significantly more opposed to tax evasion than were women. Older people were slightly more opposed to tax evasion than were younger people. Philosophy/Theology students were more opposed to tax evasion than were business/economics students.

Chapter 1: Empirical Legal Studies and Taxation in the United States §1.05

Citation	Summary
McGee, Robert W. and Geoff A. Goldman (2010). Ethics and Tax Evasion: A Survey of South African Opinion. *Proceedings of the Third Annual University of Johannesburg Faculty of Management Conference*, May 12-14.	**South Africa Graduate and Undergraduate Business Students**: Tax evasion was more justified in cases where tax funds are wasted, where the tax system is perceived as being unfair or where the government engages in human rights abuses. Women were somewhat more opposed to tax evasion than were men. Whites were significantly more opposed to tax evasion than were Africans. Catholics and Other Christians were equally opposed to tax evasion. Diploma students, undergraduate students and postgraduate students were more or less equally opposed to tax evasion. Management students and economics & finance students were equally opposed to tax evasion.
McGee, Robert W. and Susana N. Vittadini Andres (2009). "The Ethics of Tax Evasion: Case Studies of Taiwan." In Robert W. McGee (ed.), *Readings in Business Ethics* (pp. 200-228). Hyderabad, India: ICFAI University Press. An abbreviated version was published in Marjorie G. Adams and Abbass Alkhafaji (eds.), *Business Research Yearbook: Global Business Perspectives*, Volume XIV, No. 1 (pp. 34-39). Beltsville, MD: International Graphics.	**Taiwan Students**: Tax evasion was more justified in cases where there is corruption, unfairness in the system or waste (the three human rights issues were not included in this survey). Women were more opposed to tax evasion than were men.
Andres, Susana N. Vittadini and Robert W. McGee (2007). "The Ethics of Tax Evasion: A Comparative Study of Taiwan and the USA. Kaoshiung Hsien, Republic of China." Chinese Association of Political Science, September 29-30.	**Taiwan and the US Students**: US students were significantly more opposed to tax evasion than were the Taiwan students. Taiwanese women were significantly more opposed to tax evasion than were Taiwanese men. Men and women and the US sample were equally opposed to tax evasion.

Citation	Summary
McGee, Robert W. (2008). "Opinions on Tax Evasion in Thailand." In Robert W. McGee (ed.), *Taxation and Public Finance in Transition and Developing Economies* (pp. 609-620). New York: Springer.	**Thailand Undergraduate Accounting Students**: Tax evasion was more justified in cases where the government is corrupt, the tax system is perceived as being unfair or where the government wastes the tax funds collected. Women were more opposed to tax evasion than were men.
McGee, Robert W. and Serkan Benk (2011). "The Ethics of Tax Evasion: A Study of Turkish Opinion." *Journal of Balkan & Near Eastern Studies* 13(2): 249-262.	**Turkey Undergraduate Business & Economics Students**: Tax evasion was more justified in cases where the government is corrupt or wasteful or where the government engages in human rights abuses. Men were significantly more opposed to tax evasion than were women. Older people were more opposed to tax evasion than were younger people.
McGee, Robert W., Serkan Benk, Halil Yıldırım and Murat Kayıkçı (2011). "The Ethics of Tax Evasion: A Study of Turkish Tax Practitioner Opinion." *European Journal of Social Sciences* 18(3): 468-480.	**Turkey Accounting Practitioners**: Tax evasion was more justified in cases where the government is corrupt or wasteful or where the tax system is perceived as unfair. Men were significantly more opposed to tax evasion than were women. Older people were more opposed to tax evasion than were younger people.
Nasadyuk, Irina and Robert W. McGee (2007). "The Ethics of Tax Evasion: Lessons for Transitional Economies." In Greg N. Gregoriou and C. Read (eds.), *International Taxation* (pp. 291-310). Oxford: Elsevier.	**Ukraine Law Students**: Tax evasion was more justified in cases where tax rates were too high, where the government imprisons people for their political beliefs or where the tax system is perceived as being unfair. Men and women were equally opposed to tax evasion.

Citation	Summary
Nasadyuk, Irina and Robert W. McGee (2008). "The Ethics of Tax Evasion: An Empirical Study of Business and Economics Student Opinion in Ukraine." In Robert W. McGee (ed.), *Taxation and Public Finance in Transition and Developing Economies* (pp. 639-661). New York: Springer. A different version of this study that included comparative data was published under the title "Ethics and Tax Evasion in Ukraine: An Empirical and Comparative Study." *Accounting and Finance in Transition* 5: 169-198 (2008).	**Ukraine Graduate and Advanced Undergraduate Accounting and Business & Economics Students**: Tax evasion was more justified in cases where the government is corrupt, where the government imprisons people for their political beliefs or where the tax system is perceived as being unfair.
McGee, Robert W. and Silvia López Paláu (2008). "Tax Evasion and Ethics: A Comparative Study of the USA and Four Latin American Countries." In Robert W. McGee (ed.), *Taxation and Public Finance in Transition and Developing Economies* (pp. 185-224). New York: Springer.	**United States and Four Latin American Countries (Colombia, Ecuador, Puerto Rico and the Dominican Republic) Accounting and Business Students**: Tax evasion was more justified in cases where the government is wasteful or corrupt or where the tax system is perceived as being unfair. The United States was more opposed to tax evasion than was the Latin American sample in total, but Colombia was more opposed to tax evasion than were any of the other countries. US Hispanics were more opposed to tax evasion than was the total US sample. Scores for the Dominican Republic were substantially and consistently lower than for the other countries, indicating that tax evasion was less of a moral problem for the average Dominican than for the other four groups sampled. Women were more opposed to tax evasion in 17 of 18 cases. Gender differences were most significant for the three human rights arguments. Business students were more opposed to tax evasion than were accounting students.

Citation	Summary
McGee, Robert W. and Sheldon R. Smith (2007). "Ethics and Tax Evasion: A Comparative Study of Accounting and Business Student Opinion in Utah." American Society of Business and Behavioral Sciences 14th Annual Meeting, Las Vegas, February 22-25, 2007. Published in the *Proceedings of the American Society of Business and Behavioral Sciences* 14(1): 1175-1185.	**USA (Utah) Accounting and Business Students**: Tax evasion was more justified in cases where the government engaged in human rights abuses or was corrupt. Business students were slightly more opposed to tax evasion than were accounting students but the difference was not significant.
McGee, Robert W. and Sheldon R. Smith (2009). "Ethics and Tax Evasion: A Comparative Study of Utah and Florida Opinion." In Robert W. McGee (ed.), *Readings in Accounting Ethics* (pp. 343-364). Hyderabad, India: ICFAI University Press.	**USA (Utah & Florida) Accounting Students**: Tax evasion was more justified in cases where the government engages in human rights abuses or where the government is corrupt or where the tax system is perceived as unfair. Utah students were more opposed to tax evasion than were Florida students.
USA — UTAH & NEW JERSEY McGee, Robert W. and Sheldon R. Smith (2008). "Opinions on the Ethics of Tax Evasion: A Comparative Study of Utah and New Jersey." Presented at the 39th Annual Meeting of the Decision Sciences Institute, Baltimore, November 22-26, 2008. Published in the Proceedings at pp. 3981-3986.	**USA (Utah & New Jersey) Business Students**: Utah students were significantly more opposed to tax evasion in all 18 cases.

Another group of studies examined views on tax evasion using the World Values and Beliefs survey data. The findings of those studies are summarized in Table 1.3.

Table 1.3   Summary of Tax Evasion Studies Using World Values Survey Data

Asia
McGee, Robert W. (2008). "Opinions on Tax Evasion in Asia." In Robert W. McGee (ed.), *Taxation and Public Finance in Transition and Developing Economies* (pp. 309-320). New York: Springer.

Sample: 16,809 from 13 Asian countries. Women were more opposed to tax evasion in 9 of 13 countries. Older people were generally more opposed to tax evasion than were younger people.

Australia & New Zealand
Gupta, Ranjana and Robert W. McGee (2010). "Study on Tax Evasion Perceptions in Australasia." *Australian Tax Forum* 25(4): 507-534.

Sample: 3,000+ students and nonstudents in Australia and New Zealand using the same methodology as the World Values surveys.

Australia: Women were significantly more opposed to tax evasion than were men. Older people were more averse to tax evasion than were younger people. Buddhists were significantly less opposed to tax evasion than were Roman Catholics, Protestants or Orthodox Christians.

New Zealand: Women were significantly more opposed to tax evasion than were men. People over age 49 were significantly more opposed to tax evasion than were younger people. Law students were significantly more opposed to tax evasion than were accounting, business and medical majors. Faculty was significantly more opposed to tax evasion than were the student groups. Graduate students were significantly more opposed to tax evasion than were undergraduate students. Hindus were significantly more opposed to tax evasion than were other religions. Tax evasion was not considered as serious a crime as some other crimes listed.

Australia, New Zealand and the United States
McGee, Robert W. and Sanjoy Bose (2009). "The Ethics of Tax Evasion: A Comparative Study of Australian, New Zealand and USA Opinion." In Robert W. McGee, *Readings in Business Ethics* (pp. 125-142). Hyderabad, India: ICFAI University Press.

Views on tax evasion are not significantly different by country. Women were significantly more opposed to tax evasion in all three countries. Older people were more averse to tax evasion than younger people in all three countries. People in Australia and New Zealand tended to become less averse to tax evasion as the level of education increased. In the United States the difference in mean scores was not significant for the education demographic, although more educated people tended to be less averse to tax evasion.

Egypt, Iran and Jordan
McGee, Robert W. and Sanjoy Bose (2006). "Attitudes toward Tax Evasion in the Middle East: A Comparative Study of Egypt, Iran and Jordan." *Accounting and Finance in Transition* 3: 23-34.

Sample: 6,642. Egyptian women were significantly more opposed to tax evasion than Egyptian men. Gender differences were not significant for Iran and Jordan. Although older people in all three countries were more opposed to tax evasion than younger people, the difference was significant only for Jordan. Although people in all three countries tended to become less opposed to tax evasion as the level of education increased, the difference in mean scores was significant only for Jordan.

Fifteen Transition Countries and Two Developed Countries
McGee, Robert W. (2006). "Cheating on Taxes: A Comparative Study of Tax Evasion Ethics of Fifteen Transition Economies and Two Developed Economies." *Accounting and Finance in Transition* 3: 273-289.

> Women were more opposed to tax evasion in all 17 countries. In all 17 countries the oldest group (50+) was more opposed to tax evasion than the youngest group (15-29). There is a tendency to become less opposed to tax evasion as the level of education increases. Married people were more opposed to tax evasion than were single people in all 17 countries.

Germany (East & West) and the United States
McGee, Robert W., Inge Nickerson and Werner Fees (2006). "The Ethics of Tax Evasion: A Comparative Study of Germany and the United States." Working Paper, Barry University, October.

> Females were more opposed to tax evasion in all three samples. East German males were less averse to tax evasion than were males in West Germany and the United States. United States females were more opposed to tax evasion than were the females in the other two samples. People tended to become more averse to tax evasion in all three samples.

Korea, Japan and China
McGee, Robert W. (2008). "Tax Evasion, Tax Misery and Ethics: Comparative Studies of Korea, Japan and China." In Robert W. McGee (ed.), *Taxation and Public Finance in Transition and Developing Economies* (pp. 137-165). New York: Springer.

> Japan was significantly more opposed to tax evasion than China or Korea. The difference in mean scores between China and Korea was not significant. Japanese women were significantly more opposed to tax evasion than were Japanese men. Gender differences for China and Korea were not significant. Older people were more averse to tax evasion in all three countries. Opposition to tax evasion deteriorates somewhat as the level of education increases for all three countries. Buddhists were significantly more opposed to tax evasion than were Roman Catholics. None of the other comparisons of religion were significant. Married people were significantly more opposed to tax evasion than were single people.

Malaysia
Ross, Adriana M. and Robert W. McGee (2011). "A Demographic Study of Malaysian Views on the Ethics of Tax Evasion." Published in the Proceedings of the 2011 Spring International Conference of the Allied Academies, Orlando, April 6-8, 2011.

Men and women were equally opposed to tax evasion. Views did not differ by age. There is slightly less opposition to tax evasion as the level of education increases, but the difference in mean scores was not significant. Protestants were most opposed to tax evasion, followed by Roman Catholics, Muslims, Hindus and Buddhists. The group most opposed to tax evasion was the group where God was least important in their life. Housewives were most opposed to tax evasion, followed by the unemployed, full-time and self-employed, students and retired. Part-time employees were least opposed to tax evasion. The group most opposed to tax evasion was the highest income group. The group second most opposed to tax evasion was the lowest income group. There appears to be a tendency for those at higher income levels to be more opposed to tax evasion than people in the lower income groups. Divorced people were more opposed to tax evasion than were other groups. The four groups that were least opposed to tax evasion were the four groups with the most children. People who lived in cities with populations between 20,000 and 50,000 were significantly more opposed to tax evasion than were people who live in cities of other sizes. The upper class and the lower middle class had identical mean scores and were the two groups most opposed to tax evasion. There were no significant differences based on ethnicity.

Moldova and Romania

McGee, Robert W. (2009). "Views toward Tax Evasion: A Comparative Study of Moldova and Romania." *ICFAI Journal of Public Finance* 7(3&4): 7-24.

Gender did not make a difference in either country. In Moldova, people in the 30-49 group were significantly more opposed to tax evasion than people in the 15-29 and 50+ groups. In Romania, the 50+ group was significantly more opposed to tax evasion than either the 15-29 or 30-49 age groups. In Moldova, people in the upper education level were significantly more opposed to tax evasion than were people in the lower or middle education groups. In Romania, people with lower level education were somewhat more opposed to tax evasion than were people in the middle education group. Married people were significantly more opposed to tax evasion than were single people in both countries.

The Netherlands

Ross, Adriana M. and Robert W. McGee (2011). "A Demographic Study of the Netherlands Attitudes toward Tax Evasion." Published in the Proceedings of the 2011 Spring International Conference of the Allied Academies, Orlando, April 6-8, 2011, forthcoming in the *Journal of International Business Research.*

Women were significantly more opposed to tax evasion than were men. Although age made a difference in viewpoint, it was not clear what the difference was, since the pattern was not linear. In general, aversion to tax evasion tended to decrease as the level of education increased. Religion was not a significant variable. Those who attend religious services more than once a week were most opposed to tax evasion. Housewives and retired people were the two groups most opposed to tax evasion. Full-time employees and self-employed individuals were least opposed. Divorced and widowed people had equal mean scores and were the most opposed to tax evasion. In general, people with more children were more opposed to tax evasion than people with fewer children.

Poland

Ross, Adriana M. and Robert W. McGee (2011). "A Demographic Study of Polish Attitudes toward Tax Evasion." Published in the Proceedings of the 2011 Spring International Conference of the Allied Academies, Orlando, April 6-8, 2011, forthcoming in *the Academy of Accounting and Financial Studies Journal*.

The difference between male and female views was not significant. Older people were more opposed to tax evasion than were younger people. People at the lower ends of education tended to be more averse to tax evasion than people with more education. Widows and married people tended to be more opposed to tax evasion than other groups. The difference in mean scores was not significant for the variable social class. People who were very happy and quite happy were more opposed to tax evasion than were people who were not very happy and not at all happy, but the differences in mean scores were not significant.

Puerto Rico

McGee, Robert W. and Silvia López Paláu (2007). "The Ethics of Tax Evasion: Two Empirical Studies of Puerto Rican Opinion." *Journal of Applied Business and Economics* 7(3): 27-47. Reprinted in Robert W. McGee (ed.) (2009), *Readings in Accounting Ethics* (pp. 314-342). Hyderabad, India: ICFAI University Press.

Women were more opposed to tax evasion than were men. Older people were more averse to tax evasion than were younger people. People tend to become less averse to tax evasion as the level of education increases. Attitude towards tax evasion does not differ by religion.

Six Countries (Brazil, Russia, India, China, United States, Germany)

Ross, Adriana M. and Robert W. McGee (2011). "Education Level and Ethical Attitude toward Tax Evasion: A Six-Country Study." Published in the Proceedings of the 2011 Spring International Conference of the Allied Academies, Orlando, April 6-8, 2011. Forthcoming in the. *Journal of Legal, Ethical and Regulatory Issues*.

Men were somewhat more opposed to tax evasion in Brazil. Women were somewhat more opposed to tax evasion in the United States, China, India and Russia. In Germany, overall, the mean scores were identical, meaning there was no significant difference between male and female attitudes towards tax evasion. However, men with university degrees were significantly more opposed to tax evasion than women with university degrees. Older people tend to be more opposed to tax evasion than young people in Brazil, Russia, China, the United States and Germany. Age was not a significant factor for India. Regarding education, there was no clear pattern for Brazil, the United States and Germany. For Russia and China, the more education, the less opposition to tax evasion. For India, the more education, the more opposition to tax evasion.

South Africa

Ross, Adriana M. and Robert W. McGee (2011). "A Demographic Study of South African Attitudes on Tax Evasion." Published in the Proceedings of the 2011 Spring International Conference of the Allied Academies, Orlando, April 6-8, 2011, forthcoming in the *Journal of Economics and Economic Education Journal.*

Women were significantly more opposed to tax evasion. Older people tended to be more opposed to tax evasion than younger people. The most educated groups tended to be the most opposed to tax evasion. Muslims were most opposed to tax evasion, followed by Pentecostals, Evangelicals, Independent African Church members, Protestants, Roman Catholics and Jehovah Witnesses. The middle income groups tended to be more opposed to tax evasion than either the low-income or high-income groups. Those in the higher income groups were least opposed to tax evasion. Divorced people were most opposed to tax evasion; people living together as married were least opposed. Whites were most opposed to tax evasion, followed by colored (dark). South Asians and blacks were least opposed to tax evasion and had identical mean scores.

Switzerland

Ross, Adriana M. and Robert W. McGee (2011). "Attitudes toward Tax Evasion in Switzerland: A Demographic Study." Published in the Proceedings of the 2011 Spring International Conference of the Allied Academies, Orlando, April 6-8, 2011, forthcoming in the *Business Studies Journal.*

Women were significantly more opposed to tax evasion. In general, older people were more averse to tax evasion than younger people. German speakers were most opposed to tax evasion, followed by Italian and French speakers. Those with less education tended to be more averse to tax evasion than those who had more education, although the differences in means scores were not significant. Widowed was the group most opposed to tax evasion, followed closely by married and divorced. The group least opposed to tax evasion was the single/never married category. Social class was not a significant variable.

Taiwan

McGee, Robert W. and Susana N. Vittadini Andres (2009). "The Ethics of Tax Evasion: Case Studies of Taiwan." In Robert W. McGee (ed.), *Readings in Business Ethics* (pp. 200-228). Hyderabad, India: ICFAI University Press. An abbreviated version was published in Marjorie G. Adams and Abbass Alkhafaji (eds.) (2007). *Business Research Yearbook: Global Business Perspectives*, Volume XIV, No. 1 (pp. 34-39). Beltsville, MD: International Graphics.

> Men were more opposed to tax evasion. Older people tended to be more averse to tax evasion than younger people. People tend to become less averse to tax evasion as the level of education increases.

Ten Transition Countries

McGee, Robert W. and Wendy Gelman (2008). "Opinions on the Ethics of Tax Evasion: A Comparative Study of 10 Transition Economies." In Robert W. McGee (ed.), *Accounting Reform in Transition and Developing Economies* (pp. 495-508). New York: Springer.

> Women were significantly more opposed to tax evasion in five cases (Czech Republic, Estonia, Latvia, Russia and Ukraine). Women were more opposed to tax evasion in three cases but the difference was not significant (Hungary, Lithuania and Poland). Men were more opposed to tax evasion in two cases (China and Vietnam) but the differences in mean scores were not significant. In 9 of 10 cases, older people were significantly more averse to tax evasion than younger people. In the case of Vietnam, age was not a significant variable.

Ten Transition Countries

McGee, Robert W. (2008). "Changing Attitudes toward the Ethics of Tax Evasion: An Empirical Study of 10 Transition Economies." *Accounting and Finance in Transition* 5: 145-154. Also Fifth International Conference on Accounting and Finance in Transition. London, July 12-14, 2007. Reprinted in Robert W. McGee (ed.) (2008). *Taxation and Public Finance in Transition and Developing Economies* (pp. 119-136). New York: Springer, under the title "Trends in the Ethics of Tax Evasion: An Empirical Study of Ten Transition Economies."

> This study examined the trend of opinion towards tax evasion over a period of time. The people of Belarus, Estonia, Lithuania, Russia, Slovenia and East Germany became significantly less opposed to tax evasion over time. The people of Bulgaria and Poland became significantly more opposed to tax evasion over time. People in China and Latvia became somewhat more opposed to tax evasion over time but the difference was not significant.

Thailand and Vietnam

McGee, Robert W. (2006). "A Comparative Study of Tax Evasion Ethics in Thailand and Vietnam." *Journal of Accounting, Ethics & Public Policy* 6(1): 103-123.

> The Vietnamese were significantly more opposed to tax evasion than were the Thais. Women were more opposed to tax evasion in both countries, but the difference was significant only in Thailand. Older people were more averse to tax evasion than were younger people.

Thirty-Three Countries
McGee, Robert W. and Michael Tyler (2007). "Tax Evasion and Ethics: A Demographic Study of Thirty-Three Countries." *International Journal of Business, Accounting, and Finance* 1(1): 95-114.

> Women were more opposed to tax evasion in 32 countries. Opposition was equal in one country (Portugal). In almost all cases, the percentage of people who think tax evasion is always unethical rises as they get older. People are less likely to view tax evasion as always unethical if they are more educated, or at least as they move out of the lowest category of education. The percentage of people who view tax evasion as always unethical declines as income increases.

Vietnam
McGee, Robert W. (2008). "A Survey of Vietnamese Opinion on the Ethics of Tax Evasion." In Robert W. McGee (ed.), *Taxation and Public Finance in Transition and Developing Economies* (pp. 663-674). New York: Springer.

> Men were more opposed to tax evasion than women. People of all ages were equally opposed to tax evasion. People with a middle-level education are more opposed to tax evasion than are people with either a higher or lower education. People with an upper education are least opposed to tax evasion. Ancestral worshippers were most opposed to tax evasion. Roman Catholics were least opposed to tax evasion. Separated people were the most strongly opposed to tax evasion. Divorced people were least opposed to tax evasion.

§1.06 SUMMARY AND CONCLUSIONS

Although ELS is a relative new trend, empirical studies under whatever name trace their roots to the early 20th century. Traditionally, empirical legal scholars were trained in the social sciences, be it in sociology, economics, political science, finance and criminology, which led to a segmentation of empirical legal research. ELS is credited with providing for a multidisciplinary center for law-related empirical scholarship.

Empirical legal studies cover a broad field of study as should be expected, since the law itself touches almost every aspect of life and society. A joint ELS bibliographical project by UCLA and Cornell law school shows as many as 380 different research topics, varying from affirmative action, asylum and gun control to illegal aliens and, indeed, taxation. The number of tax-related studies is small with a total of 51 mentioned in the project, but given that some studies are listed in more than one sub-heading, the actual number is only 29. The bibliography is incomplete and many more empirical studies have been published, also in the area of taxation, and we have listed some of the work of the first author of this chapter as example.

Where taxes are levied with the objective to influence people's behavior, which is the case with so-called sin taxes, i.e., taxes on demerit goods and services, or with so-called green taxes that seek to discourage the use of polluting products and indirectly promote consumers to shift to the use of more environmental friendly products, the question arises whether the objectives are being realized; in other words, is the tax law successful in establishing a change in consumer behavior? A reality test should also be conducted for other legislation that encourages behavioral change, such as the mandate for airbags and safety belts in automobiles, or the prohibition of use of mobile phones and texting while driving. Policy makers need to know whether the roads are really safer as a result of these laws or whether drivers simply amend their behavior to avoid detection or safety products turn out to have unexpected negative consequences off-setting the anticipated positive results. In other words, there is a growing need for empirical legal studies to assess the effectiveness of laws as to their stated objectives. Such research can assist lawmakers in their consideration of how to change (partly) failing laws. More importantly, however, it may teach valuable lessons as to the limitations of legislation in realizing certain political goals and the need to conduct impact research prior to introducing legislation. As we have seen in the research description above, for example, a tax increase as a tool to influence consumer behavior was unsuccessfully applied to tobacco products where no reduction in consumption was realized, but would achieve the desired effect, i.e., reduced consumption, when applied to gasoline.

That is exactly the *raison d'être* of ELS: to be the reality check for past and future policies and legislation. Traditional research methods are enhanced by the employment of a solid base of up-to-date empirical information, and empirical research has the ability to help legal scholars determine whether a particular law or process is actually achieving its stated objectives. The challenge for ELS-scholars lies in raising interest for their assessments outside academia, raise the level of credibility of their research, and to become part of the policy-making process.

ELS scholars use tools and methodology generally applied in the social sciences, in particular statistical analysis. ELS-scholars offer a positive theory, which is then tested using quantitative techniques developed in the social sciences. The evidence may be produced by controlled experiment or collected systematically from real world observation. In either event, quantitative or statistical analysis is a central component of the project. As we have explained, even the more developed research designs have their limitations, and there exist reasons to be cautious with the outcome and conclusions of ELS. A critical mindset of the scholar and rigorous peer review are proven tools to avoid or discover weaknesses in scope, methodology and data gathering. Notwithstanding its structural and inherent limitations, ELS is bound to enhance and complement traditional legal scholarship, especially with the emergence of increasingly sophisticated statistical tools and more powerful computers.

§1.07 REFERENCES

Abrahams, Kenneth S. (2008). *The Liability Century: Insurance and Tort Law from the Progressive Era to 9/11*. Cambridge: Harvard University Press.

Arlen, Jennifer and Eric Talley.(2008) "Experimental Law and Economics." New York University School of Law, NYU center for Law and Economics, *Law & Economics Research Paper Series*, Working Paper No. 08-30.

Britt, Steuart Henderson (1940). "The Rules of Evidence — An Empirical Study in Psychology and Law." Cornell Law Quarterly 25: 556-580.

Brune Jr., Herbert M. and John S. Strahorn (1940). "The Court of Appeals of Maryland, a Five Year case Study." *Maryland Law Review* 4: 343-389.

Campbell, Richard V. (1933). "Statistical Survey." *Wisconsin Law Review* 9: 5-10.

Cane, Peter and Herbert M. Kritzer (eds.). (2010). *The Oxford Handbook of Empirical Legal Research*. Oxford: Oxford University Press.

Clark, Charles E. and Harry Shulman (1934). "Jury Trial in Civil Cases — A Study in Judicial Administration. *Yale Law Journal* 43: 867-885.

Clark, Charles E. and Harry Shulman (1937). *A Study of Law Administration in Connecticut: A Report of an Investigation of the Activities of Certain Trial Courts of the State*. New Haven: Yale University Press.

Clark, Charles E. and Emma Corstvet (1938). "The Lawyer and the Public: An A.A.L.S. Survey." *Yale Law Journal* 47: 1272-1293.

Committee to Study Compensation for Automobile Accidents (1932). *Report, by the Committee to Study Compensation for Automobile Accidents, to the Columbia University Council for research in Social Studies*. Philadelphia: Press of International Printing.

Davis, Sue, et al. (1993) "Voting Behavior and Gender on the U.S. Court of Appeals." *Judicature* 77: 129 et seq.

Douglas, William O. (1932). "Some Functional Aspects of Bankruptcy." *Yale Law Journal* 41: 329-364.

Douglas, William O. (1933). "Wage Earner Bankruptcies." *Yale Law Journal* 42: 591-642.

Eagleton, William L. (1932). "Academic Preparation for Admission to Law School." *Illinois Law Review* 26: 607-644.

Eisenberg, Theodore (2000). "Empirical Methods and the Law". *J. American Statistical Association* 95: 665.

Eisenberg, Theodore. "The Origins, Nature, and Promise of Empirical Legal Studies and A Response to Concerns." *Cornell Legal Studies Research Paper*, December 17, 2010.

Eisenberg, Theodore, et al. (2006). "Juries, Judges and Punitive Damages: Empirical Analyses Using the Civil Justice Survey of State Courts 1992, 1996, and 2001 Data." *Journal of Empirical Legal Studies* 3: 263-295.

Ellickson, Robert C. "Trends in Legal Scholarship: A Statistical Study, Yale University, Program for Studies in Law, Economics, and Public Policy." *Working Paper # 229*, June 1999. Revised version also published in *Journal of Legal Studies*, vol. 28, January 2000.

Epstein, Lee and Jack Knight (1997). "The New Institutionalism, Part II." *Law & Cts*, 7(2): 4.

Epstein, Lee and Gary King (2002). "The Rules of Inference." *U. Chic. L. Rev* 69(1): 1.

Everson, George (1919). "The Human Element in Justice." *Journal of the American Institute of Criminal Law and Criminology* 10: 90-99.

Feinsinger, N.P. (1932). "Observations on Judicial Administration of Divorce Law in Wisconsin." *Wisconsin Law Review* 9: 26-48.

Frankfurter, Felix and James M. Landis (1925). "The Business of the Supreme Court of the United States. A Study in the Federal Judicial System." *Harvard Law Review* 38: 1005-1059.

George, Tracey E. "An Empirical Study of Empirical Legal Scholarship: The Top Law Schools." Vanderbilt University Law School, Law and Economics, *Working Paper No. 05-20*.

Gilman, Howard (1997). "The New Institutionalism, Part I: More and Less Than Strategy: Some Advantages to Interpretative Institutionalism in the Analysis of Judicial Politics." *Law & Cts.* 7(1): 6.

Gordon, Robert W. (1993). "Lawyers, Scholars, and the "Middle Ground.'" Mich. L. Rev. 91: 2075, 2085.

Grant, John. (1929). "The Single Standard in Grading." *Columbia Law Review* 29: 920-955.

Harris, Silas A. (1933). *Appellate Courts and Appellate Procedure in Ohio*. Baltimore: John Hopkins Press.

Heise, Michael (2002). "The Past, Present, and Future of Empirical Legal Scholarship: Judicial Decision Making and the New Empiricism." 69 *U. Ill. L. Rev* 819-850.

Hines, N. William. *Empirical Research: What Should We Study and How Should We Study It?* Association of American Law Schools, http://www.aals.org/am2006/theme.html.

Kritzer, Herbert M. (2009). "Empirical Legal Studies before 1940: A Bibliographic Essay." *J. Empirical Legal Stud.* 6: 925.

Kritzer, Herbert M. "Empirical Legal Studies before 1940: A Bibliographic Essay." *Legal Studies Research Paper Series, Research Paper 09-27*, University of Minnesota Law School, August 2009.

Law David S. and David Zaring (2010). "Law versus Ideology: The Supreme Court and the Use of Legislative History." *William and Mary L. Rev.* 51(5): 1653.

Lawless, Robert M., Jennifer K. Robbennolt and Thomas S. Ulen.(2010) *Empirical Methods in Law*. New York: Aspen Publishers.

Lindgren, James (2006). "Predicting the Future of Empirical Legal Studies." *Boston University L. Rev.* 86: 1447.

Marshall, Leon C., et al. (1932). *The Divorce Court Volume One: Maryland*. Baltimore: John Hopkins Press.

Marshall, Leon C., et al. (1933). *The Divorce Court Volume Two: Ohio*. Baltimore: John Hopkins Press.

Martin, Edward M (1936). *The Role of the Bar in Electing the Bench in Chicago*. Chicago: University of Chicago Press.

Martin, Edward M. (1936). "The Selection of Judges in Chicago and the Role of the Local Bar Therein." *American Political Science Review* 30: 315-323.

McGee, Robert W. (ed.). *The Ethics of Tax Evasion in Theory and Practice*. New York: Springer (forthcoming 2012).

Miles, Thomas J. and Cass R. Sunstein (2008). "New Legal Realism." *U. Ch. L. Rev.* 75: 831.

Mott, Rodney L., Spencer D. Albright and Helen R. Semmerling (1933). "Judicial Personnel." *Annals of the American Academy of Political and Social Science* 167: 145-155.

Mott, Rodney L. (1936). "Judicial Influence." *American Political Science Review* 30: 295-315.

Murray, Frank (1938). "Requirements for Admission to Law School." *Kentucky Law Journal* 26: 290-297.

Nourse, Victoria and Gregory Shaffer (2009). "Varieties of New Legal Realism: Can a New World Order Prompt a New Legal Theory?" *Cornell L. Rev.* 95: 61.

Posner, Richard A. (1993). "What Do Judges Maximize? (The Same Thing Everybody Else Does)." *Sup. Ct. Econ. Rev.* 3(1): 1.

Rowland, C.K. and Robert A. Carp (1996). *Politics and Judgment in Federal District Courts*. Kansas: University Press of Kansas.

Rundell, Oliver. "The Time Element in Criminal Prosecutions in Wisconsin." *Bulletin of the University of Wisconsin* 512 (1912).

Schauer, Frederick (2000). "Incentives, Reputation, and the Inglorious Determinants of Judicial Behavior." *U. Cin. L. Rev.* 68: 615.

Schuck, Peter H. (1989). "Why Don't Law Professors Do More Empirical Research?" *J. Legal Educ.* 39: 323.

Slayton, Robert W. and Philip P. Brown (1939). "A Study of Pendency in Texas Civil Litigation." *Texas Law Review* 18: 1-26.

Smith, Reginald Heber (1919). *Justice and the Poor*. New York: Carnegie Endowment for the Advancement of Teaching.

Suchman, Mark (2006). "Empirical Legal Studies: Sociology of Law, or Something ELS Entirely?" *Amici* 13(2): 1-4.

Ulen, Thomas S. (2002). "A Novel Prize in Legal Science: Theory, Empirical Work, and the Scientific Method in the Study of Law." *U. Ill. L. Rev.* 875: 909-914.

Warren, Elisabeth (2002). "The Market for Data: The Changing Role of Social Sciences in Shaping the Law." The Thirteenth Thomas E. Fairchild Lecture, University of Wisconsin Law School, November 2, 2001, *U. Wisc. L. Rev.*: 1-42.

Woods, Walton J. (1916). "Necessity of Public Defender Established by Statistics." *Journal of the American Institute of Criminal Law and Criminology*: 230-244.

Warner, Sam Bass (1920). "Procedural Delays in California." *California Law Review* 8: 369-383.

PART II
Statistical Science in Tax Auditing

CHAPTER 2
Statistical Sampling in Tax Compliance Auditing in the USA

Robert Schauer

§2.01 INTRODUCTION

Sampling has been used in all audit fields for much of the 20th century in the United States. These audit fields include financial, internal, and government auditing. In tax compliance auditing, a subset of government auditing, sophisticated sampling procedures, as a direct result of advances in computer technology, are much more common. Indeed sampling can be said to be done in many ways and computers and technology have dramatically increased what is possible (or practical).

This advance in technology has impacted the auditor and how the audit is done. The auditor can design, draw, and review a sample in ways that previously were impractical. But even more fundamental, the analysis can be done in ways that allow for better audit results. How this is done depends on the reasons for sampling, and the type of auditing performed.

Audit sampling can be used for various reasons. We find sampling done for estimation purposes, discovery purposes, compliance testing, and possibly other motives. Sampling is done using objective and subjective methods, judgmental and nonjudgmental procedures, and scientific and non-scientific approaches, all of which have been in use for many years. As such, a discussion of ***statistical sampling*** (also referred to as scientific sampling) cannot be done without a discussion involving other kinds of sampling. Indeed we often find a blending of methods, sometimes obfuscating the true type of sampling.

This chapter is about sampling in tax compliance auditing in the US. In an audit, the auditor is charged with verifying whether the correct tax has been reported to the government agency for a certain period. Normally the audit scope in tax compliance audits is broad and will include many reporting periods where returns are filed over several years. The auditor can take several approaches in doing the audit. The most popular approach is to establish a total tax difference that equals the tax not paid (or possibly overpaid) on the returns. This is sometimes referred to as **tax difference auditing**. This difference may well be nothing (the taxpayer is in compliance), an amount due to the agency (a tax deficiency exists where the taxpayer is not in compliance), or depending on the nature of the tax, an amount due to the taxpayer (a tax overpayment). The auditor prefers to do a complete examination, meaning a detailed coverage of the records should be done if practical. Because of the possible enormous task of doing a complete review, especially in a large audit with a broad audit scope, some sort of sampling is commonly done in the place of a complete examination. The purpose of the sampling here is to provide an **estimate** of the total tax difference.

§2.02 WHY SAMPLE?

The basic definition of a **sample** is any part of the **population** selected for audit inspection. A population is an aggregation of business transactions or bookkeeping entries. Usually in tax compliance auditing, populations consist of economic transactions that are of audit interest (transactions falling under the scope of the examination) or elements that make up the tax return. The population, or accounting, may be available to the auditor in either paper or electronic forms.

Note that the definition of a sample does not fix any method by which the sample is taken from the population. In fact, all different kinds of sample designs are used. But regardless of the design, there are only a few basic objectives for using a sample in an audit besides the most common objective of determining the total tax difference. A basic objective could include discovery of errors if they exist and possibly determination of their frequency. Another objective could include prevention. If it is known that the auditor can use a sample, where it is possible that any part of the population could be audited, this could encourage accurate reporting as well as deter fraud. Typically the auditor will extrapolate or **project** the sample results to represent a total that would have resulted had a complete audit of the entire population (called a **census** by a statistician) been done.

In fact, the objectives for sampling can be the same as that for performing the audit itself. But what are the reasons for looking at less than 100% audit of the transactions? The most obvious reason is that a population of interest is voluminous and a complete of audit of transactions possibly would be costly in terms of both time and expense to the government agency and possibly the taxpayer. Less obvious is that in a sample, an auditor can devote *more* time to each of the sample items. If a complete audit of a large population is made, by comparison the time devoted to each population element would likely be much less, and therefore subject to more **audit error**. Audit error is any error made by the auditor whereby given sufficient audit evidence the auditor has otherwise made an incorrect conclusion. Although it usually is presumed that a complete audit would be more accurate, this assumes that audit error would be the same regardless of whether a sample or detail is done. Indeed, sampling can be used to reduce the risk of audit error.

If the auditor wishes to reduce the risk of audit error by sampling, the auditor will then assume another risk. Note that in estimation sampling, where a total value is projected from the sample results, it is used to represent the total that would have resulted had a complete examination been made *presuming* that the same general audit techniques were used. But the sample estimate is likely different than the true total. This difference between the projected total and the true value being estimated is **sampling error** and is not the same as audit error. By using a sample, even if valid and properly executed, the auditor will assume the risk of sampling error that would otherwise not be present if no sample had been done. The degree of sampling error may be small or large depending on the nature of the population, the sampling design, the size of the sample, and indeed what sample is drawn. Sophisticated sampling procedures can be used to both quantify and control this error. And this is the key to why technology and computers can help the auditor. Use of this technology puts the auditor in a position to more easily use these sophisticated procedures thereby reducing both audit error and sampling error.

§2.03   WHAT IS STATISTICAL SAMPLING?

In taking a sample, the auditor can take many different types of samples. However, all samples can be classified in two general ways. An auditor could take a **probability sample**. Any sample whereby each element of the population has a known, but not necessarily equal, chance of selection is a probability sample. Many different kinds of probability samples can be taken, and a few will be described in this document. All samples other than probability samples are **judgmental samples**. Both types of sampling are often used in tax compliance auditing, but in order to objectively estimate sampling error—a basic requirement for statistical sampling, a probability sample must be taken. By taking a probability sample, this enables the auditor to objectively measure sampling error. It is quite impossible to do this with a judgmental sample.

Probability samples can be drawn "with replacement" or "without replacement." Where a population is sampled without replacement, a unit in the population may not be selected into a sample more than once. Here, the population is repetitively reduced by one unit after the selection of each item in the sample whereby the item is "removed" from the remaining population being further sampled. In sampling with replacement, the item is not removed, and all population items can potentially be sampled more than once. Despite being more complex, sampling without replacement is preferred in practice for several reasons. First, it is slightly more efficient. That is for the same sample size, it is likely to be slightly more accurate when compared to sampling without replacement. More importantly, a sample that includes an item more than once may be difficult to explain and less agreeable to a taxpayer. Finally, although sampling without replacement is more technically complex, computer software can easily handle any additional complications.

But is use of a probability sample in an audit always considered a statistical sample? The accounting profession takes the position that in order to be a statistical sample, *two* basic requirements must be met:[1]

(i) use of a probability sample; and
(ii) use of probability theory to evaluate sample results, including measurement of sampling risk.

The second requirement, the evaluation, is usually accomplished by computing a **confidence interval**. A confidence interval is a range of values bounded by a **lower confidence limit** (LCL) and an **upper confidence limit** (UCL) derived from the sample results. The confidence intervals contains, with a specified **confidence level**, the true population value being estimated. The confidence level represents the frequency that all possible confidence intervals coming from all possible samples[2] will contain the true population amount. The confidence level can be set at different levels (commonly from 75% to 95%) depending on the requirements of the situation (but is practically set by agency policy). Construction of a confidence interval requires the auditor to apply statistical formulas, which, as can be expected, are not as straight forward as many would like to see. However, again the computer can aid in this endeavor. A detailed example of a confidence interval follows. Indeed, the example, which is actually rather simple when compared to the manner in which most sampling is actually done, illustrates why computer technology can help the auditor.

Statistical sampling in general can provide the auditor with several things. First, accuracy can be objectively measured. Second, in the long run probability samples will provide for better results when compared to judgmental samples of similar sizes. But really, the most important factor is that statistical samples can—depending on how the results are applied—perhaps provide to an auditor a legally defensible position without having to audit the entire population.[3]

In some cases, agencies might take some sort of probability sample, but not perform an evaluation. In practice such samples may be presented as "statistical samples" by virtue of the sampling method. According to the commonly accepted definition stated above, they are not.

Further, probably sampling is not the most prevalent design used by auditors. **Block sampling**, a form of judgmental sampling, is often applied when electronic records are not available, or where an agency (or an auditor) has not committed to more sophisticated methods. A block sample is drawn from a portion of the population using non probabilistic methods. This is usually a large contiguous segment of the population subjectively selected by the auditor based on convenience. Like other judgmental samples, block samples *cannot* be used to objectively measure their accuracy. Further, it is known that any projected total from a block sample likely will

---

1. International Federation of Accountants (IFAC), International Standards on Accounting (ISA) 530.
2. All possible samples of that design that are of a particular sample size and population size.
3. W. Edwards Deming, "Standards of Probability Sampling for Legal Evidence," *The American Statistician* (February 1958).

not be as accurate when compared to a projection from a probability sample of equal size.

Block sampling has a fundamental problem in one of its underlying assumptions concerning the frequency or proportion of error in the population. Block sampling presumes that this frequency or proportion is uniform throughout the population, and there is no significant difference between the sample and the part of the population not sampled. This may or may not be the case. Many populations will have error at varying amounts or frequencies caused by different factors that are not always apparent. Probability samples will likely not suffer as much as a result of this problem. Although the same phenomenon (the rate of error is materially different between the sampled and non-sampled units) can occur in the probability sample, the risk of this happening can be measured.

In the US, use of block sampling is common; however, most agencies take the position that their use is subject to taxpayer agreement. Given their inherent inability to objectively measure accuracy, and the fact that they tend to be less accurate than a probability sample, some taxpayers have successfully challenged assessments based on block samples. In these cases where successful challenges have been made, it often is the case that the non-sampled records were available but were ignored for purposes of convenience. The best that can be done with judgmental samples is a subjective evaluation of the suitability of the sample.

Subjective judgments are then routinely made (rightly or wrongly) as to whether the judgmental sample is "representative" of the population. Here the word representative or "representativeness," often found with judgmental sampling literature, is really a non-statistical term and hard to define in scientific or objective terms. It is in some cases just a euphemism for the concept of accuracy itself.[4] Although block sampling and judgmental sampling have their place in auditing, auditors often rely on these when statistical sampling should have been employed. Over-reliance on the subjective "representativeness" of the sample can easily occur. Auditor resistance to adopting more advanced methods is common—they often are never aware of how inaccurate samples have been. Often rationalized thinking takes over in the belief that these samples cannot possibly have been unsatisfactory given the frequency of their use. It should be noted that as a direct result of the revolution in technology, it is harder for those in the profession resisting change to justify their audit conclusions with less than optimal methods. The natural question can be asked: Why did the auditor use an unsophisticated sample when a scientific sample controlling risk could have easily been done?[5]

In recent years, this question and the development of technology has spurred a fresh look at sampling in audits. In many cases agencies are just now requiring their auditors to use more appropriate sampling techniques whenever possible. There has been an effort put out by many taxing agencies, in the last thirty years, to upgrade and

---

4. From a statistical perspective, a valid random sample is "representative" of the population. The accuracy of a probability sample, however, can be objectively quantified.
5. Harold Jennings and Robert Schauer, *Why Use Statistical Sampling*, MTC Review (Summer 2008), www.mtc.gov.

improve sampling efforts (post-computer era). Historically (including the pre-computer era) the auditors have always sampled in virtually all audit disciplines, including tax compliance auditing and financial auditing. But what has changed is the emergence of the computer, and the fact that records are now electronic, where they were once on paper. Tax administrators, who understand that they may have to justify their audit practices, want better sampling techniques along with efficiencies that are gained by sampling.

§2.04 SAMPLING AUTHORITY AND SAMPLING IN OTHER AUDIT FIELDS

Ultimately there is no agency or entity that judges the correctness of sampling procedures across all audit fields in the US, other than science and mathematics. Therefore reliance on statisticians, or "experts," is often crucial to the use of sampling in the US. The "battle of the experts" often results when disagreements occur. But yet that status of "expert" is not really conferred by any authority. True experts are sometimes difficult to find in that statisticians may be unfamiliar with the various types of auditing and for what purposes sampling is being used.

Financial auditors will look to the American Institute of Certified Public Accountants (AICPA) for guidance on sampling in their field of expertise. Auditors in areas outside of financial auditing may also use AICPA pronouncements to justify and support their sampling methods. Such reliance may not be entirely appropriate for the auditors not in the area of financial auditing for various reasons. While the AICPA has done a good job in guiding auditors for purposes of auditing financial statements, it would be a fallacy to assume the appropriate or even best procedures used for financial auditors will apply equally well in other audit areas. Indeed the sampling standards of the AICPA have come under fire in that—in the opinion of some—the "level of proof" supplied by non-statistical procedures has been put "on the same par" as statistical sampling.[6] The criticism is that this has proven to be detrimental to those relying on the work of financial auditors. Because this paper concerns itself with sampling in tax compliance, whether or not financial auditors follow the prescribed guidelines, or whether those guidelines are flawed with respect to the "level of proof" argument, is not explored further here. However, it can be said, that the guidelines set out by the AICPA for statistical sampling designs used by financial auditors in their audits are well thought out, and much of those have been "borrowed" into other areas of auditing in the US.

Financial auditors are generally concerned with the correctness or reasonableness of the company financial statements. Internal auditors will be concerned with compliance with company policies and rules as well as a sundry of other roles. Government tax auditors include federal, state, and local government agencies auditing taxpayers for income taxes, excise taxes, sales and use taxes, property taxes, and other taxes. The sampling methods for these three groups often overlap and are shared,

---

6. Neal B Hitzig, "Statistical Sampling Revisited," *The CPA Journal* (2004).

Chapter 2: Statistical Sampling in Tax Compliance Auditing in the USA   §2.04

but there are distinctions and dangers in applying methods between these areas of auditing, especially when it comes to sampling. Experts sometimes are not conscious of the fundamental differences that separate these fields.

Historically, statistical sampling was probably first most prevalent in financial auditing, and was used to some degree from the 1950s forward. This auditing tool has since expanded into other areas. For example, the US Internal Revenue Service (IRS) started applying statistical sampling on a regular basis in their tax audits in the 1970s.[7] State agencies auditing for sales and use tax adapted the financial sampling, and particular the statistical sampling as used by the IRS, in their audits beginning generally in the 1980s (and somewhat before in larger states) and even further in 1990s. The expansion, adoption, and oversight of statistical sampling has been sporadic, and some state agencies have not fully adopted these techniques to this date and rely wholly on judgmental sampling.[8]

When statistical methods were first applied in the US to auditing, statistical experts came from the field of survey sampling. Audit sampling is sometimes seen in the US as a subset of survey sampling. Indeed, an audit of business transactions can be thought of as a "survey" in the sense that the auditor will be concerned with a "finite population." Many of the authoritative references regularly used in audit sampling actually are sources that are focused on survey sampling.[9] Audit sampling borrows heavily on survey sampling and employs sample designs first used in surveys, including **simple random sampling** (more detail follows), **stratified sampling** (also more on this later), **cluster sampling**,[10] and **multistage sampling**.[11]

Financial auditors apply statistical sampling using diverse methods. They may want to sample for determining whether any errors likely exist (**discovery sampling** or sometimes referred to as **exploratory sampling**), whether material error within an

---

7. *Sampling Methods for the Auditor; an Advanced Treatment* (Arkin), pages 100-109.
8. Federation of Tax Administrators Task Force on EDI Audit and Legal Issues for Tax Administration, *Appendix A: Survey of State Sampling Practices*, (January 2004), www.taxadmin.org/fta/pub/Samp2004.pdf.
9. Survey sampling references:

    - William Cochran, *Sampling Techniques*, John Wiley and Sons (1977).
    - William Edwards Deming, *Some Theory of Sampling*, Dover Publications, Inc (1950).
    - Morris H Hansen, William N Hurwitz, and William G. Madow, *Sample Survey Methods and Theory, Volumes #1 and #2*, John Wiley & Sons (1993).
    - Leslie Kish, Survey Sampling, John Wiley & Sons (1995).
    - Paul Levy and Stanley Lemeshow, *Sampling of Populations: Methods and Applications*, 3rd Edition, John Wiley & Sons, New York, (1999).
    - *Sampling: Design and Analysis*, Sharon Lohr, Duxbury Press, (1999).
    - Richard Scheaffer, William Mendenhall, R. Lyman Ott, *Elementary Survey Sampling*, 5th edition, Duxbury Press (1996).

10. Cluster sampling is similar to stratified sampling in that the population is divided into groups, or clusters. Unlike stratified sampling, the group itself or cluster is the sampling unit whereby a simple random sample is used to select the clusters. Those clusters chosen in the sample are then audited in their entirety.
11. Multistage sampling is similar to a cluster sample, but instead of auditing the selected clusters in their entirety. The clusters are then referred to as primary or first-stage units. The selected clusters are sub-sampled. The audit is performed on the sub-sampled units (called secondary or second-stage units).

internal control system exists (***compliance testing***), or whether the account balances are reasonably stated (***substantive testing***). In discovery sampling, financial auditors will often take a simple random sample and make an acceptance or rejection criteria applied to the population as a whole. If no errors are found the population "is accepted." Otherwise, if one or more errors are found in the sample, the population "is rejected" and other audit or sampling are done. For compliance testing, that is whether the audited client is satisfactorily adhering to a set of internal controls generally applied to accounting data, a wide variety and sometimes complex procedures can be found in practice. For a thorough discussion of this topic, a variety of publications are available to explain these methods.[12] Finally, another commonly used sampling design is estimation sampling. Estimation sampling is used to determine some value or frequency.

If the financial auditor desires information of "how many" or at what frequency some unknown quality exists, the auditor will likely apply an ***attributes*** measurement. For example, the auditor concludes after performing a statistical sample of the population of interest, with a 95% confidence level, that the rate of error is less than 1% in the entries into the book ledgers.

If the auditor wishes to determine a value, such as an average, a total, or an account balance, the auditor will apply something called ***variables*** measurement. For example, the auditor concludes after performing a valid statistical sampling procedure, with 95% confidence that the true total accounts receivable is between USD 1.5 million and USD 1.7 million, with the most likely value being USD 1.6 million. Here the USD 1.5 million is the lower confidence limit (LCL) and the USD 1.7 million is the upper confidence limit (UCL). The most likely value, USD 1.6 million, is often referred to as the ***mid-point*** or ***point estimate***.[13] If the recorded balance is USD 1.65 million, the auditor might accept that as being substantially correct because it is within the confidence interval, and the margin of difference between the recorded value and the point estimate is deemed immaterial (if USD 0.05 million can indeed be deemed

---

12. Sampling sources for financial auditors:

    - AICPA SAS 39—*Audit Guide*.
    - Auditing Practice Release; *Audit Sampling*, AICPA, (1999).
    - Alvin Arens & James K Loebbecke, *Applications of Statistical Sampling to Auditing*, Prentice Hall, Englewood CA (1981).
    - Alvin Arens & James K Loebbecke, *Auditing: An Integrated Approach*, Prentice Hall, Englewood CA (1994).
    - Herbert Arkin, *Sampling Methods for the Auditor; an Advanced Treatment* (1982).
    - Herbert Arkin, *Handbook of Sampling for Auditing and Accounting*, McGraw Hill, New York, 1974.
    - Dan M Guy, *An Introduction to Statistical Sampling in Auditing*, John Wiley and Sons, 1981.
    - Donald A Leslie, Albert D Teitlebaum, & Rodney J. Anderson *Dollar-unit Sampling, A practical Guide for Auditors*, CCH (1979).
    - TW McRae, *Statistical Sampling for Audit and Control*, John Wiley & Sons, London, 1974.
    - Maurice S Newman, *Accounting Estimates by Computer Sampling*, John Wiley & Sons, New York (1982).
    - Donald A Roberts, Statistical Auditing, AICPA (1978).

13. This is an estimate of some ***parameter*** of the population.

immaterial). If there a substantial deviation with the point estimate, the auditor may request a book entry adjusting the accounts receivable balance to that amount. This is called a *point estimate adjustment*. On the other hand, if the recorded balance was less than the lower confidence limit of USD 1.5 million, the auditor may request a correcting entry to that amount. Alternatively if it is more than the upper confidence limit of USD 1.7 million, the auditor may request an entry adjusting the balance down to that limit. These corrections to either the LCL or UCL are often referred to as *interval adjustments*. Point estimate adjustments and interval adjustments are also used in tax compliance auditing.

In financial auditing there are two distinct ways of performing variables sampling. The traditional methodology is sometimes referred to as **classic transaction sampling**, which will be explained in more detail later in this text. Classical transaction sampling is also the method of choice for most tax compliance auditing, particularly for the IRS and state taxing agencies. Auditors can take either a simple random sample or a stratified random sample using the classic approach. In a simple random sample, all units will have an equal chance of being selected into the sample, irrespective of its book value.[14] In stratified random sampling, whereby sampling units are placed in separate groups according to an attribute, the overall chance of selection into the sample across all units is not equal, but is a function of this attribute. In nearly all stratified populations, the attribute used to stratify is a book value. This has the effect of weighting the sample according to the book value of the population units. Units with a greater book value will be given a higher chance of being selected. Indeed a common practice of auditors and accountants is to give more attention to larger dollar value transactions or records. An alternative approach often used in the financial audits for variables sampling is sometimes referred to as **sampling proportion to size** (PPS sampling).[15] Similar to PPS is **monetary unit sampling** (also known as dollar unit sampling).[16] PPS sampling in the US is restricted mostly to financial auditing and is rarely used in tax compliance auditing for a variety of perceived disadvantages.[17] PPS sampling has some distinct advantages over traditional transaction sampling in that it can be easily adapted to other types of sampling, including attributes sampling making

---

14. This is the commonly applied definition. It is more technically correct to say that in a simple random sample, every sample of the same size from a given population has an equal probability of being selected.
15. PPS sampling is an alternative to stratified sampling, but still gives a greater chance of selection to units with a larger book value. In PPS sampling, the chance of selection into the sample for any unit is directly related to the size of its book value relative to the total book value for the entire population. In a properly executed PPS sample, done with replacement, units from the population may be represented in the sample more than once.
16. Monetary unit sampling (MUS) differs from PPS sampling as it defines each dollar as a sampling unit, where each dollar has an equal chance of selection. Practically the effect is similar to PPS sampling. Although technically a MUS sample can be selected without replacement, population items can appear in the sample more than once. For example, a USD 100 transaction represents 100 different sampling units. Each dollar has an independent chance of being in the sample. Therefore transactions can appear in the sample more than once.
17. A major concern is that PPS or MUS might have transactions appearing more than once in the sample. While technically valid, explaining this result to unsophisticated taxpayers may become problematic. This problem is avoided in simple random sampling or stratified random sampling.

it attractive to financial auditors. Since most taxing agencies are much more concerned over the total value of the tax errors, rather than their frequency, attributes sampling is also rarely used in tax compliance auditing in the US.[18]

§2.05 SAMPLING THEORY: SIMPLE RANDOM SAMPLING EXAMPLE

Before discussing sampling theory and methods by which to draw these samples, an example will be given. In order to discuss the example, statistical formulae will be used. Statistical formulae are an expression of the "language" of statistics. It should be noted, while some notations are common, statisticians typically use their own style, and as result major differences exist in the literature. The notations used here are in line with Cochran's *Sampling Techniques*. In addition, for the purposes of the example below, it is assumed that the reader understands some basic concepts, including the following:

(1) **Mean value** (arithmetic mean value in a group of numbers).
(2) **Modal value** (most frequently occurring value).
(3) **Standard deviation** (a measure of variability of a group of numbers in relation to the arithmetic mean).
(4) **Frequency distribution** (tabular expression of a group of numbers).

---

18. Sampling sources focused on tax compliance auditing:
    - *Audit Manual, Chapter 13, Statistical Sampling; Sales and Use Tax*, California State Board of Equalization (January 2000), www.boe.ca.gov/pdf/fam-13.pdf.
    - *Sampling for Sales and Use Tax Compliance*, Federation of Tax Administrators (December 2002), http://www.taxadmin.org/fta/pub/sample.pdf.
    - *Auditing in an Electronic Environment (e-auditing) and Stratified Statistical Sampling*, Florida Department of Revenue (2002), booklet #GT-300034, dor.myflorida.com/dor/forms/2002/gt300034.pdf.
    - *RAT-STATS Companion Manual*, Department of Health & Human Services, OIG—Office of Audit Services (September 2001), oig.hhs.gov/organization/oas/ratstats/CompManual2007.pdf.
    - *RAT-STATS 2007 User Guide, version 2*, Department of Health and Human Services, Office of Inspector General, Office of Audit Services (Revised October 2004), oig.hhs.gov/organization/oas/ratstats/UserGuide2007.pdf
    - IRS Training Publication: *Basic Statistical* Sampling, Training 3172-001 (Revised August 1993) TPDS 871251.
    - IRS Training Publication: *Advanced Statistical Sampling*, Training 3174-002 (Revised May 1992) TPD3 87030A.
    - *Sampling Policy & Guideline Manual*, Multistate Tax Commission (July 2008), www.mtc.gov/Audit.aspx?id=612.
    - *Computer Assisted Audits, Guidelines and Procedures for Sales Tax Audits*, New York State Department of Taxation & Finance (October 2001), Publication 132, www.tax.state.ny.us/pdf/publications/sales/pub132_1001.pdf.
    - *Statistical Sampling for Sales and Use Tax Audits*, Tennessee Department of Revenue, Audit Division, Revised (April 1999), www.state.tn.us/revenue/tntaxes/sales/statsamplingapr06.pdf.
    - *Statistical Sampling Manual*, Washington State Department of Revenue (revised January 2008).
    - Will Yancey, *Statistical Sampling in Sales and Use Tax Audits*, CCH (2002).

(5) **Normal distribution** (or "Gaussian distribution," a probability distribution that has a bell shape curve).

In the example the auditor is asked to review a tax return of a generic unspecified tax. To arrive at the tax, a tax measure is computed by deducting an amount from a gross figure. The tax rate applied to the tax measure is 10%. Specifically, the auditor wishes to verify the validity of the deductions taken from the gross figure. If any deductions are found to be invalid, that is in error, an assessment of 10% of tax will be made on those deductions. The audit objective is to determine the total deducted in error (taxable error). The individual deductions range from a **book value** of USD 50 to USD 2000.[19] The total of all the deductions found on the supporting documentation agrees with the deduction amount taken from the gross to arrive at the tax measure. The auditor notes that there is a population of 1500 deductions, whereby **N = 1500**. The auditor will establish an **error value** on each deduction denoted with a value y. If there is no error in the amount deducted, $y = 0$. On the other hand, given the facts of the example, if error is noted in any deduction, a positive error value will be recorded up to or equal to the book value. Since the population consists of 1500 units, respective error values for each population unit is signified by y_i. The total error value for the population, **Y**, will be determined as follows, $Y = y_1 + y_2 + y_3 + ... + y_N$. Another expression of this is $Y = \sum_{i=1}^{N} y_i$. Note that Y does not represent the total book value of all deductions and if the taxpayer is in complete compliance, $Y = 0$.

But a complete review would be burdensome for both the auditor and the taxpayer. It was therefore decided that a probability sample will be performed because the auditor wishes to objectively measure sampling error. The largest sample practical under the circumstances was deemed to be 100, where **n = 100**. Because the auditor is now sampling, an *estimate* of the total error value, $\hat{Y}$, will be made from the sample results.

The auditor is also further constrained by another important fact. In compiling the deduction amount, the taxpayer kept a running tally of the deduction in paper form. Therefore, an electronic file of the deductions is unavailable. This is an important consideration in planning the sample. Because such a file is unavailable, it is impractical to use advanced sampling techniques that would most likely improve the accuracy of the sample results. In most cases, the auditor would make further refinements to the population and then likely use a stratified sampling design in drawing the sample. Under the circumstances, the auditor decided to use a simple random sample, drawn without replacement. In the absence of the electronic file, much of what could be done by the computer now must be done through more labor intensive processes.

To draw the sample, the auditor uses a sampling software package. A sample of 100 random numbers will be taken, such that any possible random number in the

---

19. In *Sampling Techniques* (Cochran) the book value is described simply as a form of an **auxiliary variate** (explained at page 150). The assumption in accounting is that this book value is somehow related to, or **correlated** with the variable being estimated, which in tax compliance auditing is most often an error value. In fact the utility of the book value is usually present only if some correlation exists between the two values.

sample has **correspondence** to only one of the 1500 **sampling units**. The correspondence, or matching of the random numbers to the sampling units, will in this case be done mechanically by hand (with the computer file, this task could be performed by the computer software). The software generates a set of "random" numbers using a **pseudo-random number generator** (PRNG).[20] One of the hallmarks of such a procedure is the generation of a *seed*, allowing set of random numbers representing the selected sample units to be reproduced or expanded. Computers have allowed great efficiencies in the process of sampling that previously was done manually. Today, the entire sampling process is automated and documented by the computer. By mechanizing the process, the risk of procedural error is greatly reduced.

The auditor, once the specific deductions are identified on the listing through the process of correspondence, hands a list of needed documents to the taxpayer for the 100 sampled deductions. The taxpayer then later provides the auditor with these documents, and they are examined for their deductibility. An error value is recorded on each sample item y_1 through y_{100}. The auditor finds 92 of the sample items have no error. But eight of the deductions are declared invalid, and an error amount is recorded for each of these:

y_{14}= 1559.00	y_{50}= 1995.00	y_{75}= 1361.00	y_{84}= 1450.00
y_{20}= 350.00	y_{51}= 643.00	y_{77}= 433.00	y_{85}= 167.00

For any other sample items, $y_i = 0$ (the entire audited sample is included as an appendix). Using these sample error values, a mean error value, $\bar{y}$, is computed for the sample as follows: $\bar{y} = (y_1 + y_2 + y_3 + ... + y_{100})/n$, which can also be expressed as $\bar{y} = \left(\sum_{i=1}^{n} y_i\right)/n$. In this case $\bar{y} = 79.58$.

Note that if the auditor did a complete examination rather than a sample, the mean population error value (or true mean), $\bar{Y}$, would be $\bar{Y} = (y_1 + y_2 + y_3 + ... + y_{1500})/N$ (where the y_i items represent units from the population, not the sample). The true total error value, Y, can be computed as $Y = N\bar{Y}$. To find an estimate of total error, $\hat{Y}$, using the **mean-per-unit estimator**,[21] the auditor simply substitutes the population mean, $\bar{Y}$,

---

20. A PRNG is an algorithm used to generate a sequence of numbers that approximate the properties of random numbers. The numbers supplied are a result of a deterministic process and are not truly random as they are predictable, but have qualities of randomness, and are therefore referred to as *pseudorandom*. Eventually a PRNG starts repeating the same sequence of numbers. The **period** is the maximum length of the sequence of numbers before a PRNG begins to repeat. PRNGs have different periods and some exhibit better qualities of randomness than others. There are tests that can be run against these algorithms to test their suitability. It is widely accepted that for purposes of estimation sampling, a good PRNG with an adequate period will be sufficient for the purposes of producing random samples.
21. The IRS and other agencies will actually refer to this estimator as the **difference estimator** as the tax differences are being estimated directly. Further, what the IRS (and other agencies) refer to as the mean-per-unit estimator is altogether something else, and incorporates the book values in the computation. This unfortunately is a cause for some confusion. The definition and formulas for these estimators, as used by the IRS and other agencies can be found in various publications referenced at the end of this chapter.

with the sample mean, $\bar{y}$, such that $\hat{Y} = N\bar{y}$. The auditor works out the equation as (1500) (79.58) = 119,370 as an estimate of the total error. It is also interesting to note here that the auditor, with this estimator, really made no use of the book value in deriving the estimated total error.

Using the point estimate of USD 119,370 taxable error computed, the taxpayer could be sent a tax assessment of USD 11,937 (excluding any applicable interest and penalty). But how good of an estimate is this point estimate? To answer this question, the auditor could construct a confidence interval around the point estimate.

(To underscore a point discussed earlier, some auditors do not construct the interval and proceed directly with the point estimate, but then also maintain that because a simple random sample was performed, a "statistical sample" was done. This, at best, is a questionable statement. Because the auditor failed to construct a confidence interval on the audit results, that this procedure, according to the commonly accepted definition, is definitely not a statistical sample.)

Given the limited space available, it would likely be a bit over-ambitious of a goal to provide a reader a complete understanding as to how a confidence interval is constructed. However, it is hoped that an appreciation will be gained for the work normally done by the computer. Here, the formulas are worked out manually.

To create a confidence interval, an estimate of variation that surrounds the point estimate needs to be made. To explain how this is done, some theoretical concepts need to be presented. The point estimate can be thought of as *only one* of the possible point estimates coming from all possible samples with a sample size n, out of a population size N. If all possible point estimates are computed from all possible samples, it would be possible to place them into a frequency distribution, called a **sampling distribution**. This distribution might have characteristics that the auditor could exploit. A sampling distribution is derived from a sampled population, but should not be confused with the underlying population. Further, a measurement of the variation of the sampling distribution, called the **standard error**, could be computed from this theoretical construct. The standard error is simply the standard deviation of all the observations contained within the sampling distribution. If the sampling distribution is **normally distributed**, approximately 68% of all the observations, in this case point estimates from all possible samples n out of a population of N, would be within one standard deviation of the center of the distribution. Further, other probabilities could be used, such as 95%. If the distribution is normally distributed, approximately 95% of all observations will be within a factor of about two standard deviations. One could use other probabilities as well and adjust the standard deviation factor from the standard normal table. For any given probability a certain percentage of all observations will fall some distance from the center of the distribution by some factor of the standard deviation.

Therefore, if the sampling distribution at hand is normally distributed, there is approximately a 68% probability that the point estimate computed from the sample is within one standard error, whatever that might be, of the center of the sampling distribution. Two more key pieces of information can be exploited, without having to actually compute the sampling distribution:

(1) The center, or modal value of the sampling distribution for mean-per-unit estimator, whatever it might be, exactly equals the true unknown amount.
(2) An estimate of the standard error of the sampling distribution can be computed from the sample results.

Using this information and a standard normal table, the auditor can construct a confidence interval *provided* that the auditor can assume that the underlying sampling distribution is normally distributed. For this example, it is assumed that it is normally distributed. Later, a brief discussion is given to when an auditor can and cannot assume that the underlying sampling distribution is normally distributed.

To get an estimate of the standard error, $s_{\hat{Y}}$, the auditor must first compute the standard deviation of the sample error values, s_y. This is a measurement of how the various sample error values are distributed about the sample mean, $\bar{y}$, and can be computed as $s_y = \sqrt{\sum_{i=1}^{n}(y_i - \bar{y})^2 / (n-1)}$. This formula happens to be readily available in most spreadsheet formulas available today.[22] Using the sample results, $s_y = 325.41$ (rounded to the 2nd decimal value).

Using the standard deviation computed from the sample, the standard error, $s_{\hat{Y}}$, can be computed directly from the sample results using the following formula, $s_{\hat{Y}} = N s_y \, fpc / \sqrt{n}$. Here the *fpc*, or the **finite population correction**, is computed as $\sqrt{1 - n/N}$, and is required because the sample was drawn without replacement. The estimated standard error is computed here to be $s_{\hat{Y}} = 47,156$ (rounded to the nearest whole number).

In this case, the confidence level that is required by the agency is a 90% confidence level. The auditor could use a factor from the standard normal table. However, since this is a simple random sample drawn from a **finite** population, it is more appropriate to obtain the factor from **Student's t-distribution**. With a sample size of 100, the **degrees of freedom** are computed to be 99. Using this basic information, the auditor can go to a **t-table** (based on the *student's t*) [23], or utilize functions available in any of today's computer spreadsheets or statistical packages.[24] Here the factor will be expressed as *t*, and has the value of 1.660391.

With all the various components of the confidence interval computed, the auditor can compute the confidence interval according to the following formula:[25]

$$\hat{Y} \pm (t s_{\hat{Y}})$$

---

22. In Excel, this formula is "=stdev(*range containing all sample values, including zero sample error values*)."
23. Sometimes, rather than use the student's t, auditors will use the z-score from the standard normal distribution.
24. In Excel, this formula is "=TINV(*area in the tails where the probability is 2-sided, degrees of freedom*)."
25. See *Sampling Techniques* (Cochran) at page 27 (formula 2.24).

Chapter 2: Statistical Sampling in Tax Compliance Auditing in the USA §2.05

The first term, $\hat{Y}$, is the point estimate of the total error and was previously computed to be USD 119,370. The second term, $ts_{\hat{y}}$, subtracted and added from the point estimate, is an estimate of the sampling error[26] using a 2-sided confidence level of 90%, and is $ts_{\hat{y}} = (1.660391)(47,156)$, or 78,297 (rounded to the nearest whole value). Subtracting and adding the estimate of the sampling error, we get the lower confidence limit (LCL) of USD 41,073 and upper confidence limit (UCL) of USD 197,667. Expressing this in a statistical statement, the auditor can be 90% confident that the true unknown total error is between USD 41,073 and USD 197,667.

But how is this statement useful to the auditor? Note that **relative precision**, a measurement of the comparative accuracy of the sample, can be expressed in percentage form as $(sampling\ error/point\ estimate)$, or in this case, 66%. Is this a sufficiently accurate sample to justifying an assessment based on the point estimate of USD 119,370? This, of course, is subject to things like agency policy, and possibly other precedents. Ordinarily, most people are comfortable with relative precision, or accuracy, of 5% or better (if accuracy is better than 5%, that means that relative precision has a numerical percentage less than 5%). Some jurisdictions have relaxed the standard where relative precision can be as much as 70% in order to justify an assessment based on the point estimate.

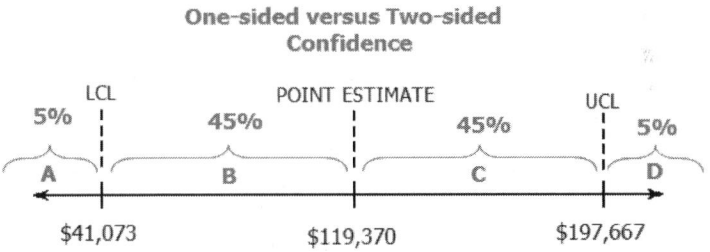

A closer look at the confidence interval using the example as depicted in the figure above shows that the true unknown total error could rest anywhere on the arrowed line (whereby the true total must be under A, B, C, or D). After taking the sample, the auditor can only be 50% confident (A + B) that the true unknown is less than the point estimate, and a corresponding 50% confidence that it is greater (C + D). But using the statistical calculations of the confidence interval, in a 2-sided confidence statement, the auditor can be 90% confident (B + C) that the true unknown is between the LCL and UCL. Using the same calculations, the auditor, in a 1-sided statement, can be 95% confident (B + C + D) that the true unknown is greater than the LCL. Similarly, the auditor could be 95% confident that the true unknown is no more than the UCL (confidence here is equal to A + B + C).

Also, it should be noted that **risk** and confidence are compliments. If the auditor is 95% confident (B + C + D) that the true unknown total error is at least the LCL, then the auditor can also say that there is a 5% risk (A) that the true unknown amount is less than the LCL. Similarly, with a 90% confidence (B + C) that the true unknown total is between the LCL and UCL, that there is a corresponding 10% risk (A + D) that true

---

26. Sometimes this is referred to as the **precision amount**.

unknown total error is either less than the LCL, or greater than the UCL. Finally, it should be noted that these risk/confidence statements are with respect to sampling error. A similar objective analysis of audit error cannot be done. However it is often a relatively safe assumption that audit error will likely be less in a sample when compared to a detail, as the auditor will likely devote more time in examining each of the individual units.

Because populations in tax compliance auditing tend to be highly variable, some agencies use an interval adjustment upon which to base the assessment to deal with the problem of a wide confidence interval (where relative precision is worse than desired). Because the confidence interval is too wide, using the point estimate as a basis for the assessment becomes less reasonable (or possibly less defensible). In this case, an interval assessment could be based on the LCL of USD 41,073. When using the LCL to make an adjustment, since there is only concern of where the true unknown is in respect to that amount, it can be said that the auditor is 95% confident that the true unknown total is at least USD 41,073. In that case, the agency could send out an assessment for USD 4,107 rather than USD 11,937.

§2.06 DEALING WITH TAX COMPLIANCE POPULATIONS WITH LARGE VARIANCES

The above example is typical, in that the confidence intervals derived from the samples tend to be very large. What causes the variability in the tax compliance populations? Some think this has to do with the wide variance in the book values. However, the variation in the book values is not the main causal factor in nearly all cases. The reasons generally lie elsewhere.

Usually, tax compliance populations are mostly in compliance. Having a population of mostly zero values, and only a few nonzero values will likely cause the population to be more variable, more difficult to sample, and result in comparatively wider confidence intervals (imprecise samples). In fact, as the frequency of the zero values in the population goes up, a greater relative width in the confidence interval can be expected. In other words, the more compliant the taxpayer, the more difficult it is to sample any total error with a degree of relative accuracy desired. A thorough examination of the cause is beyond the scope of this document. But the root of the problem can be shown by dissecting the standard deviation calculation: $s_y = \sqrt{\left(\sum_{i=1}^{n}(y_i - \bar{y})^2\right) / (n-1)}$. Where zero values occur in greater frequency, provided at least some nonzero value exist in the population, the greater the degree of dispersion relative to the mean value. Introducing negative error values (tax overpayments) to the population sampled will even have a more dramatic impact (when positive error values, or tax underpayment, exist as well).

To further illustrate this problem, the results of the example can be used and modified for several different assumptions. In the example, 8 out of the 100 sample items were in error, resulting in relative precision of 66% (it is presumed that an examination of the total *population* would have roughly 8% of sampling units in error whereby about 92% of the population will have a zero value). To show the effect of possible **error rates**[27] in the population, the sample is reevaluated at different *sample error rates*. For example, for a 7% sample error rate, only the first seven nonzero sample items are accepted as errors, and the eighth is valued at zero. For 6% sample error rate, only the first six are accepted as errors, and so on:

Sample Error Rate	Rel. Prc.	$\hat{Y}$	LCL	UCL	Sampling Error	Std Error
8%	66%	119,370	41,073	197,667	78,297	47,156
7%	67%	116,865	38,573	195,157	78,292	47,153
6%	74%	95,115	24,264	165,966	70,851	42,671
5%	79%	88,620	18,324	158,916	70,296	42,337
4%	92%	68,205	5,424	130,986	62,781	37,811
3%	104%	58,560	(2,524)	119,644	61,084	36,789
2%	134%	28,635	(9,728)	66,998	38,363	23,105
1%	160%	23,385	(14,127)	60,897	37,512	22,592

Note that although the sampling error decreased with lower error rates, the relative precision (accuracy) became worse. With the lower error rates, sampling error exceeded the point estimate of the total taxable error, $\hat{Y}$. The estimate of the accuracy of the sample became worse as the error rate decreased (compliance got better). This should illustrate the problem of using the point estimate for a basis of an audit adjustment when the error rate is actually low. Unfortunately, some agencies continue to use the point estimate as a basis of the adjustment, even when the estimate of accuracy of the sample projection is very poor (where that is certainly subjective) or if relative precision exceeds 100% (by any definition, an inaccurate projection).

As can be seen in the example and evaluation at different error rates, in tax compliance auditing, there is concern or discussion about "low error rates" and the inherent problems they cause. This is the same discussion that is referred to as the "rare error" problem, as found in the financial accounting literature.[28] Here, it can also be said that if the error rate exceeded 8% in the sample results, that the relative precision would be tend to be better, that is it is more likely that a precise estimate relative to its estimated sampling error will result.

Given that the auditor wants to get a more precise result, what can the auditor do? Here the key is to understanding the problems from a statistical perspective, and exploiting technology to help achieve a better result. Observations with respect to four factors are discussed here. The last one being key to the auditor.

Note that the estimated sampling error is computed as $ts_{\hat{Y}}$, having a term on the left tied directly to the desired confidence, and the standard error term on the right

---

27. An error rate is a reference to the frequency of nonzero errors that exist in the population, usually expressed as percentage. The error in the population is a reference to a dollar amount, and usually refers to the total value of the error and will not be expressed as a percentage. There can be a population error rate (frequency of the nonzero errors in the population) and a sample error rate (frequency of the nonzero errors in the sample).
28. A. Rashad Abdel-khalik and Ira Solomon (editors), *1984 Auditing Research Symposium*, Office of Accounting Research, University of Illinois at Urbana/Champaign (1985), pages 17-19.

representing the estimated variability of the sampling distribution. The larger the sampling error, the wider the confidence interval. By reducing the term on the left, that is the confidence factor, the overall sampling error can be reduced relative to the point estimate. By simply reducing the confidence level better relative precision can be attained. Some agencies use lower confidence levels for this reason. However, this usually is not a decision within the control of the auditor, as it is normally set by agency policy. Further, at a certain point, too low of a confidence level (or too high a risk) will be deemed unacceptable making the utility of the statement of little value.

In computing sampling error, the term on the right, the standard error, is computed as follows in a simple random sample, $s_{\bar{y}} = N s_y fpc / \sqrt{n}$. By increasing sample size, n, the denominator in the formula, this will reduce the standard error and then also the sampling error, resulting in better relative precision. In the example, it should be noted that the auditor already took a sample that was considered as large as practically possible. It is usually impractical, with tax compliance populations, to control sampling error through an increase in sample size, n, in order to achieve better relative precision. Sample size may have to be set at around 5000 or higher when sample error rates around 2% or so are expected. Such sample sizes are going to be in nearly all cases far too large and impractical.[29]

Another way to reduce the standard error is to employ a stratified sampling design. Here is where technology is a key. In most cases (but not always), using a stratified sample can cut the standard error in half, or even more when compared to a simple random sample of equal size. In the example, stratification on the deduction amount was not practical, as the electronic file of deduction amounts was not available to the auditor. This then illustrates the importance of obtaining electronic files in audits.[30] To stratify, the auditor needs to have some means by which to stratify. Usually this is based on some book value. Stratification is usually only practical to the auditor with computers and software packages especially customized for these procedures (see tax compliance software packages at the end notes).[31] Also, it should be noted, that in most cases, the auditor will not use a simple random sample, but a stratified random sample, if a statistical sample is needed in the audit.

But the key method to achieving better samples is to deal directly with the problem shown above with the respect to the percentage of zeros in the tax error

---

29. A. Rashad Abdel-khalik and Ira Solomon (editors), *1984 Auditing Research Symposium*, Office of Accounting Research, University of Illinois at Urbana/Champaign (1985), page 18.
30. Frank Olken, *Random Sampling from Databases*, University of California at Berkley (1993), page 6-7.
31. Software used in tax compliance auditing with sampling capability (packages with a "*" are designed specifically to provide statistical evaluations for samples as typically used in tax compliance auditing):

    - ACL, www.acl.com (commercial software).
    - EZ-Quant Statistical Analysis Software, www.dcaa.mil/ezquant.htm (free download).
    - IDEA Data Analysis Software, www.audimation.com (commercial software).
    - MTC Sampling Software*, www.mtc.gov (software provided to anyone attending the MTC's statistical sampling training class—there is a fee charged for the training).
    - RAT-STATS*, oig.hhs.gov/organization/oas/ratstats.asp (free download).
    - Sesame Software, www.sesamesoftware.com (commercial software).
    - TSEP*, www.kulpandassoc.com (formerly commercial software, now available for free download at www.mtc.gov).

population. Regardless of sampling design, it is best if the auditor does some up-front audit planning and adjustments to the population *prior* to sampling by making efforts to reduce the number of zeros in the population. But how can this be done?

§2.07 PLANNING THE SAMPLE: POPULATION REFINEMENTS AND ESTABLISHING THE FRAME

The importance of pre-planning cannot be overemphasized in statistical sampling. To illustrate this, the concepts of book value and error value need to be explored. Auditors ordinarily will use book values to plan the sample. A book value is merely an amount from the accounting records usually readily available prior to sampling. Accounting populations are often characterized as having a **right skew** (many transactions of low dollar value, tailing off to the right with larger values in decreasing quantities):

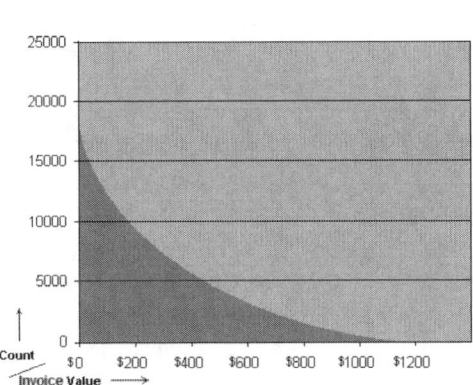

Populations could be left-tailed where the tail extends in the opposite direction, but this is atypical for most accounting populations.

Finally, as seen in text books, there is the normal distribution with classic bell-shaped curve showing no skew:

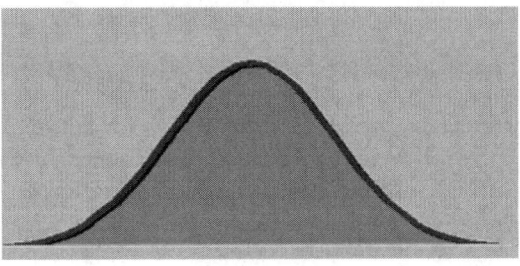

Accounting populations usually never come close in appearance to a normally distributed population and are often highly skewed. However, normal populations are nevertheless exploited by the auditor—at least in a theoretical sense. We have learned that the accompanying sampling distributions—which are hopefully normally distributed—will be used in the construction of the confidence interval. It is worth mentioning that the mean and modal values occur at the center of the normal distribution.

On a side-note, auditors sometimes provide statistical data or evaluations based on the book values used in planning. That is they attempt to satisfy the requirement of a statistical evaluation of the sample results by placing a confidence interval on the book values from the sample, while forgoing an evaluation on the value being estimated. While there is nothing wrong with placing confidence intervals based on the sample book values (sometimes there are legitimate reasons for doing so), it is again a dubious practice to claim to have done statistical sampling by doing an evaluation on the book values alone. Confidence intervals computed using the book values from the sample are usually not as wide as confidence intervals using the error values. Even of more concern, the error values usually display even greater skew. This is because the error value population is a composite or "mixture" of several different populations.

But what does the error population look like? Even as an unknown, the auditor has some information about it before sampling.

Part of the error population, the nonzero errors, often look like the underlying book values—but these values are usually only a small part of the population (as taxpayers tend to be mostly in compliance). The nonzero errors are accompanied by a much larger number of zero values, which if graphed might look like the following:

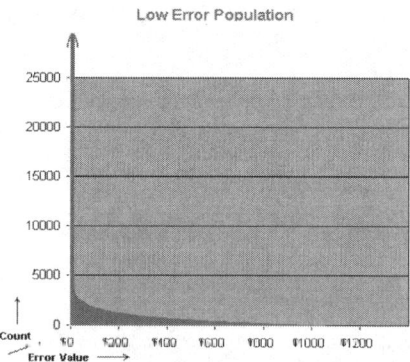

Visually, the extreme spike of zero values is interesting as the modal value. The mean value will be to the right of the spike and not at zero (compare this characteristic with normal distribution). It is also interesting to observe that using statistical measurements for skew, this population will likely exhibit much more skew than a possible refined population with many of the zero values removed. This graph could very well depict the underlying population in the example used in the previous section.

Sometimes the error populations are blends of yet another variety, one of zeros, nonzero values that are positive (underpayments of tax), *and* nonzero values that are negative (overpayments of tax). In that case, the graphed values will also extend off in the other direction to the left of the modal value of zero.

Herein lies the problem with the assumption of a normally distributed sampling distribution. If the underlying error population is skewed, a normally distributed sampling distribution will be present *only* if the sample is sufficiently large.[32] That is to say, with small samples, the sampling distribution will be nowhere near normal. Therefore, using the classical approach the auditor is unable to produce a reliable confidence interval when the sample size is deficient. If the sample is not "statistically large," the mathematical relationships of normal distribution, as explained in the previous section, will simply not be present. Therefore, the auditor must be careful, when dealing with highly skewed populations that the sample is sufficiently large. As can be expected, as the error rate decreases, typically the sample size needed will be larger to satisfy this assumption. Therefore the size of the sample has bearing on two different concerns, first the accuracy of the sample, and then whether it is sufficient in order to assume that the underlying sampling distribution is shaped (approximately) like the normal distribution.[33] The "validity of the normal assumption" provides yet another reason for planning and refining populations.[34]

When can the auditor assume that the sample is statistically large? This will be addressed later. For now, it should be noted that the zero values in the population cause statistical and accuracy problems. Dealing with the zeros in the planning stage, in the development of the population to be sampled, is the best option.

Putting aside theoretical considerations, we know that a large number of zeros (nonerrors) exist in the populations examined because it is true that most taxpayers are in compliance, and tax errors will be noted in only a much smaller number of transactions, if at all. The auditor cannot ethically turn a zero into a nonzero. But, if a zero value exists, it is not imperative that the auditor include the item in the population sampled. It should be noted that where zero values and nonzero values exist in audit situation, if the auditor can remove only zero values from the population before sampling, the total error, Y, will still be the same. Therefore, removing a substantial number of zero items before sampling means:

---

32. In actual fact, even when there is a large sample, the sampling distribution, when taken from a skewed population, will contain some skew. Here, we say that is normal or "nearly normal" in such a way that we can rely on the confidence interval. Refer to A. Rashad Abdel-khalik and Ira Solomon (editors), *1984 Auditing Research Symposium*, Office of Accounting Research, University of Illinois at Urbana/Champaign (1985), page 18.
33. Note that the assumption that the sample is statistically large, allowing the auditor to safely assume that the underlying sampling distribution is large, is a different concept than that of accuracy of the estimator. Those new to these concepts often have difficulty in distinguishing these two ideas. The usefulness of the sample has to do with the width of the confidence interval. Whether the sample itself is statistically large goes to whether one can rely on the interval as calculated, regardless of its relative width. It could be, that the sample is indeed statistically large where the interval is valid, but it is much too wide to be useful. This result can easily happen in tax compliance auditing, especially in poorly planned samples or when the auditor does not (or cannot) make efforts to refine the population.
34. Reference *Sampling Techniques* (Cochran) at pages 39-44.

(1) the total of the error remains unchanged;
(2) the *error rate* in sample will likely increase and improve the accuracy of the sample.

(After reading the second point, some may think that "increasing the error rate of the sample" will increase the assessment—not true. The same error total exists, but now in a smaller population, one without so many zeros.)

Consequently it follows that investigating the population of interest before sampling (i.e., planning the process of sampling), and obtaining as much advance information beforehand, can offer opportunities to refine sampled populations to exclude zeros. Taking time up front and carefully examining what information is available, in an electronic audit, will often be extremely beneficial to the end result. Opportunity to remove zeros from a frame not in electronic form is limited, underscoring the importance of the electronic file to the auditor.

Zeros are any transaction or unit where no error exists. These non-errors could be in records that are potentially within the auditor's scope of interest, or in records that are outside this scope. It is the latter that can be removed automatically. In the former, records should only be removed only if it is known before sampling that no error exists with a fair degree of certainty (what that "degree" is depends on the situation at hand, and the auditor's knowledge and possible experience).

This can be further illustrated by the scope and objective of the audit. The auditor will have a reason for the audit. That reason will likely mean that only a portion of the records of the taxpayer will be of audit interest. Only a certain portion of the records will contain error as defined by the audit scope. Removing those records that can never contain any error *is the best thing the auditor can do to improve the precision of the sample.*[35] For example, a sales tax auditor will get a download of sales made by a company that has many different locations. Many of these locations are in other jurisdictions, and any error made in those sales will be outside the scope of the audit. Removing these from the population and not examining them will not have any effect on the audit if all the remaining invoices were audited in detail. The same would be true if the items were removed and a sample was made of the remaining sales (of course, the sample results can be applied only to the population sampled).

In another example, a group of transactions are within the scope of the audit in that tax is applicable to those transactions. However it is known that tax is due *and* that tax has been paid correctly on those transactions. No tax error exists in these units. These can also be safely removed from the sampled population as they would otherwise represent zero values if drawn into the sample.

Unfortunately, there are some, both in government and outside it, who do not understand and/or adequately appreciate the concept explained in the above paragraphs. Some unfair criticism—most likely due to a lack of understanding—has been levied against this valid tax compliance auditing technique. As a result, some outside government have actually advocated withholding vital and relevant records from

---

35. A very good discussion of this process that (hopefully) involves both the taxpayer and the auditor is found in *Statistical Sampling in Sales and Use Tax Audits* (Yancey) at pages 22-25.

auditors for use in planning the sample.[36] On the other hand, it is also true that some auditors have inappropriately refined populations in way that slanted the audit result in the favor the government by removing items from the population where there is a likelihood that the taxpayer has paid tax in error (actually removed nonzero amounts, in some cases credit errors, from the population).

Either of these situations is unfortunate. Both the government and the taxpayer have a vested interest in determining what records require an examination. Withholding records or taking a detached view of the audit is likely not in the taxpayer's best interest. For the government's part, it needs to make sure that the auditors make proper decisions of what populations are examined (do not advocate that auditors take a blind eye toward tax overpayments). Further government should insure that their auditors utilize the information at hand. In fact "lazy auditing" is known to happen—planning or refining is sometimes not done because this involves more work and possibly special technology skills.

Another misconception is that the auditor, in statistical sampling, should not use judgment anywhere in the process. That is, the auditor should not make subjective decisions. This is not the case. Although the sample selection and evaluation are done as a result of an objective process, the use of judgment is crucial, particularly in the planning process. Development of the sampled process can be done using intuition and experience. But once the sampling frame is created, sample should be selected using probability theory.

Related to this discussion of planning and defining samples using electronic methods, it should be noted that in the last 25 years, a new specialist in computer auditing has evolved. Such a person is sometimes referred to as a **computer audit specialist** (CAS). A CAS has a technical skill set that is much more diverse than most auditors. Usually, they are trained in specialty software packages[37] that are used to supply electronic data to the audit staff for the purpose of auditing which may or may not include sampling.[38] Technical knowledge of computer data, accounting systems and software, and IT systems is needed to effectively perform the job. The CAS, as a function of the position, usually has a better knowledge of sampling then most auditors. As auditing has evolved into the 21st century, particularly on larger and more involved audits, the lone auditor is becoming less common. Tasks of communicating with the IT and other technical staff are left to the CAS, rather than the lead auditor for obvious reasons. When sampling is required, usually the work of creating the sampled

---

36. Rocky Cummings, *Too Much Information*, Tax Trends, CCH (2007).
37. Other software packages used to convert data, refine, or edit data (excluding common spreadsheet and database software):

    - ACL, www.acl.com (commercial software).
    - IDEA Data Analysis Software, www.audimation.com (commercial software).
    - Monarch, www.datawatch.com (commercial software).
    - Sesame Software, www.sesamesoftware.com (commercial software).
    - Ultra-Edit, www.ultraedit.com (commercial software).
    - VEdit, www.vedit.com (commercial software).

38. If the audit can be done efficiently without sampling, this is preferred.

populations falls onto the CAS, who consults with the auditor with the responsibility of making decisions as to what records are within the scope of the audit and other key areas regarding the direction of the audit.

Usually, a download of records is received in an electronic audit that requires some editing and other work to prepare them for auditing and sampling. The download has to be sufficiently broad to cover the accounting records and transactions covered by the scope of the audit. In some cases, the breadth of the download must necessarily be beyond the scope of the audit, to insure that all necessary records are provided to the auditor. It is also required that the auditor validate the data for audit purposes. How well this population can be refined to match with the auditor's scope has to do with the quality and sufficiency of the data received. The proper data fields and records must be present to allow the auditor to make optimal decision concerning audit procedures. For example, in the prior example, a sales tax auditor was auditing a company that did business in several different locations. To safely remove the sales that pertain to other jurisdictions, this information must be within the data provided to the auditor.[39]

Insofar as practical, the auditor should remove records that are not needed or that are known not be in error. Once that process is done, the auditor is ready to prepare the population for sampling. In doing so, the auditor is constructing a framework, or **sampling frame** (or simply the "frame"), upon which the sample will be carried out. A sampling frame is merely a device or other construct that allows the auditor to match the random numbers obtained as a result of the sampling process to the actual sampling units drawn into the sample. In a computer audit, the frame is going to be a database or listing of sample units remaining from the original download after refinements. A frame can be thought of as the **physical representation** (another term used for "sampling frame") of the population that is intended to be sampled.[40] It should also be noted, that to qualify as a properly executed statistical sampling procedure, that once the auditor has removed records from the sampling frame, the estimate and confidence interval from the sample must be applied only to the records that had a chance at being sampled. The auditor should either ignore these removed records altogether, or perform a separate examination on any records (possibly in another sample). It is not proper in a statistical sense to extend the sample results to such records.[41]

---

39. A very good discussion of the various problems found in computer data files in *Statistical Sampling in Sales and Use Tax Audits* (Yancey), at pages 45-65.
40. *Auditing Practice Release; Audit Sampling*, AICPA, (1999), page 36.
41. Sometimes the results of the sample are extended to those items removed from the sample frame prior to sampling in any case. The taxpayer may even agree to this for the sake of convenience. Even when done with mutual consent, it would be incorrect for the auditor to assert that the confidence interval applies, in a statistical sense, on any projection to those removed items that had a chance of selection into the sample. That is, no objective statement can be made about accuracy concerning a projection to any items that did not have a chance at being selected into the sample.

§2.08 PROJECTING THE SAMPLE RESULTS

In the example above, the sample results were projected using mean-per-unit estimation. [42] In fact, more than one estimator can be used to estimate the total tax difference, $\hat{Y}$. To review, the mean-per-unit estimator relies solely on the error values, y_i, in the sample to project total tax error. Note that the total taxable difference for the entire population of N units is Y, where $Y = y_1 + y_2 + y_3 + ... + y_N$. However, for each item in the population, another value, a book value, x_i, which is a *known value*, can be used along with the y_i values to estimate total tax difference. The total book value is X, where $X = x_1 + x_2 + x_3 + ... + x_N$. X will be known before sampling in most cases, and the auditor can use that total along with y_i and x_i values in the sample to project the unknown total tax difference, $\hat{Y}$, using other estimators that exploits this other information.

Each estimator will have a different point estimate of the sample, $\hat{Y}$, and the estimated standard error, $s_{\hat{Y}}$ to compute the confidence interval $\hat{Y} \pm (ts_{\hat{Y}})$. Therefore, each estimator will have its own confidence interval. In today's computerized world of auditing, software packages can easily and automatically calculate these intervals.

In practice, four estimators can be found.[43] They include the mean-per-unit, difference, ratio, and regression estimators. The example is shown below and provides

---

42. The sample results for the example are as follows (all amounts in USD):

	Book Value (x_i)	Error Value (y_i)	Audited Value ($x_i - y_i$)
Total from the sample	88,673	7,958	80,715
Mean Value from the sample	886.73	79.58	807.15
Standard deviation of values	605.71	325.41	625.10

Other information regarding the population and sample:

Sampling units in the population	(N)	1,500
Number of units sampled	(n)	100
Total Book Value of the Population	(X)	USD 1,213,786
student's t	t	1.660391
(from normal table)	z	1.644854

43. In addition to the tradition methods commonly used to project and evaluate sample results, new methods are theoretically now possible. These new methods, although relatively simple in their basic logic, exploit the computing ability of today's computers to carry out their many iterations that would be otherwise be difficult in a manual setting. These new procedures include the **Bootstrap** and the **Empirical Likelihood**. For more on these new methods, which to date are not being applied by taxing agencies, reference *An Introduction to the Bootstrap*, Bradley Efron and Robert J. Tibshirani, Chapman & Hall/CRC, 1993 and *Empirical Likelihood*, Art B Owen, Chapman & Hall/CRC, 2001. How these methods could be applied are explored in two articles found in the *Journal of Government and Financial Management*, "The Bootstrap, What the Government Auditor Should Know," (Fall 2002) and "Is Your Agency Too Conservative? Deriving More Reliable Confidence Intervals,"(Summer 2005), both by Alan Kvanli and Robert Schauer.

the point estimate ($\hat{Y}$), standard deviation ($s_{\hat{Y}}$), standard error ($s_{\hat{Y}}$), and confidence intervals (using the same assumptions as given in the example, whereby the confidence level will be set at 90%) for each estimator. The following table summarizes the respective computations for each estimator using the facts from the example (note that the confidence interval is computed using a 90% 2-sided confidence level, bounded by the upper confidence limit and lower confidence limit):

	Point Estimate	Standard Deviation	Standard Error	Sampling Error $t(s_{\hat{Y}})$ or $z(s_{\hat{Y}})$	Upper Confidence Limit (UCL)	Lower Confidence Limit (LCL)	Relative Precision
	($\hat{Y}$)	*	($s_{\hat{Y}}$)	$z(s_{\hat{Y}})$	(UCL)	(LCL)	
Mean-per-unit	119,370	325.41	47,156	78,297	197,667	41,073	66%
Difference	3,061	625.10	90,586	150,408	153,469	(147,347)	4914%
Ratio	108,931	318.57	46,165	75,935	184,866	32,996	70%
Regression	106,369	319.91	46,359	76,254	182,623	30,115	72%

* $s_{(x-y)}$ for the difference estimator; for the other three estimators: s_y

Some jurisdictions, according to a set sampling policy, will use only one estimator at all times. Other jurisdictions will possibly use one of any four estimators at any time (they will have complex rules that dictate what estimator is used, usually those rules, in essence, dictate that the estimator to be used is the one that provides the most precise result[44]). Also, software packages make certain assumptions that can make dramatic differences, particularly how the confidence coefficient is applied. Below are screenshots of the example statistically evaluated several software packages (IRS Software and Multistate Tax Commission Software).

The following is an evaluation of example using the DOS version of the IRS Sampling software:

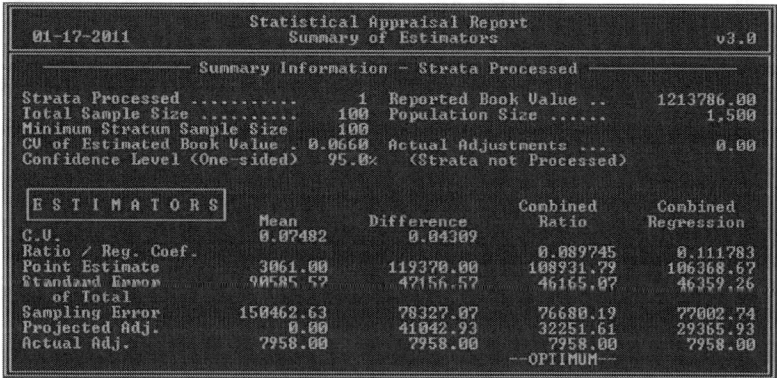

---

44. At §425 of the *Sampling Policy & Guideline Manual* of the Multistate Tax Commission is an example of these rules.

Chapter 2: Statistical Sampling in Tax Compliance Auditing in the USA §2.08

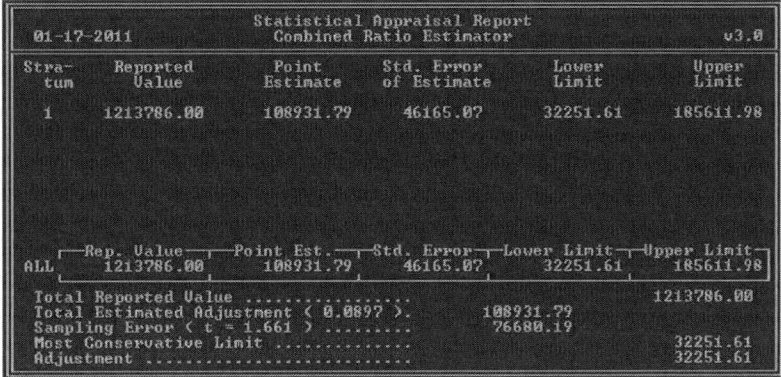

The point estimates and standard error calculations agree with the computations shown above. Note, however, that the sampling error differs from the estimates given above. This is because the IRS software uses a t-value of 1.661 for all estimators in computing sampling error, where the t-value used in the example is 1.660391 for mean-per-unit and difference estimation and the z-value of 1.644854 for ratio and regression.

Also worth mentioning is the IRS reverses the names of the mean-per-unit and difference estimators and refers to the book values as reported values or reported book values. The IRS would in this case use the ratio estimator to project total taxable error as that estimator has the lowest (estimated) standard error of USD 41,165.07. Further, the IRS would likely make an interval adjustment and use the lower confidence limit of USD 32,251.61. Other agencies would use the point estimate or compute a lower confidence limit at a different confidence level (the IRS uses a 1-sided 95% confidence level—a very conservative position). There are other restrictions that the IRS applies to sample projections that do not apply here. Under these restrictions the ratio and regression estimators are not available for projection if certain conditions are not met, even though they provide the lowest sampling error amounts, as there is concern over estimator bias for ratio and regression. In this case, the bias of ratio and regression estimators is estimated to be small, and therefore can safely be ignored.

Below is a similar evaluation of the example facts in the Multistate Tax Commission Sampling Software package:

Note that the confidence level is put at 90% 2-sided, which will provide for similar upper and lower confidence limits as computed in the IRS software that used a 95% 1-sided confidence level (the differences once again can be attributed to the fact the MTC software uses—in this case—a t-value of 1.660391 for all estimators). Another small difference terminology can be seen in that sampling error is referred to as the **Precision Amount**. Also, like the IRS software, the Multistate Tax Commission software reverses the names of the mean-per-unit and difference estimators.

Example					Ratio and Regression Statistically Available based on CV tests:		Yes
		Confidence Level:	90% (2-Sided)				
			Unstratified Evaluation Estimated Total Taxable Error *(Excluding Detail Error*)*				
Stratum Processed	Estimator	Point Estimate *	Standard Error	Precision Amount	Lower Confidence Limit *	Upper Confidence Limit *	Relative Precision Excluding Detail Error
1st Stratum Only	Mean	3,061	90,586	150,407	(147,346)	153,468	4914%
1st Stratum Only	Difference	119,370	47,157	78,298	41,072	197,668	66%
1st Stratum Only	Ratio	108,932	46,165	76,652	32,280	185,584	70%
1st Stratum Only	Regression	106,369	46,359	76,975	29,394	183,343	72%

Mean-per-unit Estimator:

Label	Symbol	Formula
Mean Taxable Error from the Sample	$\bar{y}$	$\dfrac{\sum_{i=1}^{n} y_i}{n}$
Estimated Total Taxable Error	$\hat{Y}_m$	$N\bar{y}$
Standard Deviation	$s_{y,m}$	$\sqrt{\dfrac{\sum_{i=1}^{n}(y_i - \bar{y})^2}{n-1}}$
Standard Error	$s_{\hat{Y},m}$	$\dfrac{N s_{y,m} \, fpc}{\sqrt{n}}$
Confidence Interval Estimated Total Taxable Error		$\hat{Y}_m \pm \left(t * s_{\hat{Y},m}\right)$
Sampling fraction	f	n/N
Finite Population Correction	fpc	$\sqrt{1-f}$
Student's t	t	[normal deviate]

Chapter 2: Statistical Sampling in Tax Compliance Auditing in the USA §2.08

Label	Symbol	Formula
z-score	z	[normal deviate]

This method was shown in the example, and from that it can be seen that the mean of the sample, $\bar{y}$, was used to represent the average of the population, and then this was then used to estimate the total of the population. Some jurisdictions rely solely on this estimator for several reasons. First, when compared to the other common estimators, it is more simple to apply and usually performs as well or nearly as well as the other estimators. Another reason is that the mean-per-unit estimator is an **unbiased** estimator. That is it has an attribute that was mentioned previously in that the center of the underlying sampling distribution will equal the unknown being estimated—this is not necessarily true for the other estimators. Some agencies want to avoid any mention of "bias" in their procedures, and therefore use only this method to project sample results.

Difference Estimator:

Label	Symbol	Formula
Mean Book Value from the Sample	$\bar{x}$	$\dfrac{\sum_{i=1}^{n} x_i}{n}$
Audit Value from the sample		$x_i - y_i$
Mean Audited Value from the Sample		$\bar{x} - \bar{y}$
Estimated Total Taxable Error	$\hat{Y}_d$	$X - N(\bar{x} - \bar{y})$
Standard Deviation (of audited values)	$S_{(x-y),d}$	$\sqrt{\dfrac{\sum_{i=1}^{n}[(x_i - y_i) - (\bar{x} - \bar{y})]^2}{n-1}}$
Standard Error	$S_{\hat{Y},d}$	$\dfrac{N s_{(x-y),d}\, fpc}{\sqrt{n}}$

Label	Symbol	Formula
Confidence Interval Estimated Total Taxable Error		$\hat{Y}_d \pm (ts_{\hat{Y},d})$

The difference estimator is computed in tax compliance auditing in some jurisdictions, but is rarely used as it performs poorly in nearly all populations examined. The difference estimator, rather than computing the average mean error value of the sample, computes something altogether different, the **mean audited value**. It actually is an extension of a type of estimation used in financial auditing, where it is more suitably applied.[45] The difference estimator is also an unbiased estimator of the true unknown amount.

Of further note, the naming of this estimator and the mean-per-unit estimator is often switched by some agencies. What is called herein as the mean-per-unit estimator is referred to as the difference estimator by the IRS, Multistate Tax Commission, and many of the state taxing agencies. Also, what is called here as the difference estimator will be called the mean-per-unit estimator by these agencies. Whenever reference is made to these two estimators, one should be careful and compare them to the formulas given herein or by the reference material given by the agency to identify the estimator.

Ratio Estimator:

Label	Symbol	Formula
Estimate of the Ratio of the Taxable Error	$\hat{R}$	$\bar{y}/\bar{x}$
Estimated Total Taxable Error	$\hat{Y}_r$	$\hat{R}X$
Standard Deviation	$s_{y,r}$	$\sqrt{\dfrac{\sum_{i=1}^{n}(y_i - \hat{R}x_i)^2}{n-1}}$
Standard Error	$s_{\hat{Y},r}$	$\dfrac{Ns_{y,r}fpc}{\sqrt{n}}$

---

45. The estimator is explained in a financial audit application in *Accounting Estimates by Computer Sampling* (Newman), pages 23-25. There is also an explanation of when the estimator is expected to perform well at page 77 of *Sampling: Design and Analysis* (Lohr).

Label	Symbol	Formula
Confidence Interval Estimated Total Taxable Error		$\hat{Y}_r \pm (z * s_{\hat{Y}_r})$ (some jurisdictions compute t for this estimator, and use that instead of z)

In the population, there is a ratio of errors (often called "error percentage" by auditors). The common ratio used represents the total taxable error, Y, divided by the total book value, X, found in the population expressed by the ratio R. So we can see that $Y/X = R$, and therefore $RX = Y$. X is usually known, and if R can be estimated, we can get another estimate of Y. The ratio R can be estimated from the sample and is designated as $\hat{R}$. It is computed by taking mean error value of the sample divided by mean book value of the sample: $\hat{R} = \bar{y}/\bar{x}$. The ratio $\hat{R}$ is applied to the total book value, X to estimate total error, $\hat{Y}$, and is common in statistical sampling applications. It is also the most common method of sample projection in judgmental sampling.

This estimator performs well, and often gives a slightly more precise estimate when compared to the mean-per-unit estimator. However, the center of underlying sampling distribution for this estimator does not equal the true unknown amount—the estimator is said to have bias. Usually if the auditor is careful and has a sufficiently large sample, bias is usually small and can be safely be ignored. But, because the estimator has bias, some agencies will avoid its use even if bias is small, to avoid any questions altogether.[46]

Regression Estimator:

Label	Symbol	Formula
Regression Coefficient of the Taxable Error Values	b	$\dfrac{\sum[(x_i - \bar{x})(y_i - \bar{y})]}{\sum(x_i - \bar{x})^2}$
Estimated Total Taxable Error	$\hat{Y}_g$	$N\bar{y} + b(X - N\bar{x})$
Standard Deviation	$s_{y,g}$	$\sqrt{\dfrac{\sum_{i=1}^{n}[(y_i - \bar{y}) - b(x_i - \bar{x})]^2}{n-2}}$

---

46. A discussion of this bias of the ratio estimator can be found in *Sampling Techniques* (Cochran) at pages 160-162 and *Sampling: Design and Analysis* (Lohr) at pages 66-71.

Label	Symbol	Formula
Standard Error	$s_{\hat{Y},g}$	$\dfrac{Ns_{y,g}\,fpc}{\sqrt{n}}$
Confidence Interval Estimated Total Taxable Error		$\hat{Y}_g \pm \left(z * s_{\hat{Y},g}\right)$ (some jurisdictions compute t for this estimator, and use that instead of z)

This estimator is more complex in the other three described above, and historically, has been avoided by government agencies because of these complexities. Indeed, computers make the use of the estimator relatively easy. However, explaining or describing its formula is still arduous. Because this estimator will usually provide slightly better results, more agencies are beginning to use it in audits. The estimator takes and combines all the information used by the other three estimators to make better estimates.[47]

The regression estimator uses the linear relationship of the taxable error values, y_i, with the book values in the sample, x_i, along with the total book value, X, to estimate total taxable error. To accomplish this, a regression coefficient, b, is computed from the sample results. The regression coefficient is in other applications, described as the **slope** in linear regression.[48]

The regression estimator, like the ratio estimator, is also biased in that the center of the underlying sampling distribution for this estimator does not equal the true unknown being estimated. Again, as with the ratio estimator, if the auditor is careful and takes a sufficiently large sample, this bias can be safely ignored.[49]

§2.09   STRATIFICATION

Stratifying involves dividing a population, or sampling frame, into non-overlapping groups, or strata. A simple random sample is taken independently within each stratum. Stratifying is done using one or more criteria in an effort to make the individual sampling units within a stratum more alike than they would otherwise be in an unstratified frame. More than one sample is usually taken from various strata. However, in some cases, the auditor may stratify and take only one sample from one

---

47. An interesting mathematical discussion of how the regression estimator can be considered the "parent form" of the other three estimator is included in *Accounting Estimates by Computer Sampling* (Newman) at pages 27-30.
48. In fact, Microsoft Excel includes the regression coefficient within its suite of formulas as "=slope()."
49. A discussion of this bias of the ratio estimator can be found in *Sampling Techniques* (Cochran) at pages 198-199 and *Sampling: Design and Analysis* (Lohr) at page 74.

of the groups. In any stratified sampling frame where only one stratum is sampled, it should be evaluated as a simple random sample. Otherwise, where more than one independent sample is taken from the strata, special formulas for the various estimators, including the mean-per-unit estimator, need to be applied to stratified frames. These formulas are more complex when compared to simple random sampling. Computer software is usually used to compute evaluations of stratified samples. This is an area that has benefited greatly by advance in technology. Evaluation of stratified frames can now be done with relative ease.

The primary reason for stratifying the population is to control variation and increase achieved precision in the sample evaluation. That is for a sample of equal size, the thought is that a stratified sample will outperform a simple random sample (i.e., be more precise). So if the standard error of a frame that is not stratified is designated $S_{\hat{Y}\,srs}$, and the standard error from the same population that is stratified is designated $S_{\hat{Y}\,stratified}$, stratifying will yield better results if $S_{\hat{Y}\,srs} > S_{\hat{Y}\,stratified}$.

Earlier in this text, a discussion of the normal assumption was given. Basically, the evaluation formulas can be relied on only if a sufficiently large sample was taken, so that the auditor can safely presume that the underlying sampling distribution is normally distributed. Another reason for taking a stratified random sample is that the underlying sampling distribution for tax compliance populations will likely begin to look normally distributed at smaller sample sizes when compared to simple random sampling.

It is not true that a stratified random sample is always better than a simple random sample with regard to precision. Whether a stratified random sample does better depends on several factors. By placing the units in strata where the differences between the units is small, that is the variation within the strata is less when compared to the overall population, stratifying can be expected to achieve gains. Also the greater variation between the strata, the more efficient a stratified sample will be when compared to simple random sample of the entire sampling frame. In most cases in tax compliance, the best method (and often the only way) to accomplish this is by using book values (x_i) as means by which to stratify (more on this below). The theory is that the larger tax errors will result in the strata where the book values are larger. This is usually true for the nonzero tax errors ($y_i \neq 0$). But if one takes in consideration in tax compliance that most tax errors will have a value of zero ($y_i = 0$) in that no tax error is present in the majority of sampling units, the zero tax differences will have no correlation with book values (the formula for measuring the correlation of the book values and the tax errors are in the table below). [50] The mixed population of y_i values of zero or nonzero values within the strata will likely create a sampling frame, if created on the book values x_i, where there is at least some differences in overall variation across the strata. Therefore some gains will be realized by stratifying in the majority of populations encountered in the field of tax compliance.

The number of sampled strata in a sampling frame is designated by L. By adding more strata, that is by making L larger, gains in overall precision can be realized. However the law of diminishing returns usually kicks in early, in that for most populations in tax compliance, stratifying and taking sample in more than six strata does not realize much, if any benefit.[51] In fact, in practice, most agencies will construct sampling frames that have three or four sampled strata. Some agencies do stratify more than this, but it must be said that there is usually no loss of precision by constructing more than six strata. The cost of stratifying is usually negligible, and usually an automated process.

A common practice is to create a "high dollar" stratum where no sampling takes place. The units within the high dollar stratum have the largest book values. Within this stratum, every unit is examined for tax error. In the final evaluation, the projection will exclude this stratum, and be limited to only those strata sampled. However, the final outcome of the audit will consider both the projection from the sampled strata and the actual amount of error found in the high dollar stratum.

The book value that demarks the high dollar stratum is sometimes referred to as a *ceiling*. Any sampling unit where the book value is above the ceiling will be in the high dollar stratum. By placing a ceiling on the sampling frame and auditing those units on an actual basis, sampling error is reduced. In other words, there are reasons from a theoretical point of view for establishing a ceiling amount and auditing some part of the population on an actual basis. From a practical point of view, it should stand to reason that the units with the largest book value will likely impact the overall total error by the largest degree. It is a common practice in accounting to separately review the largest transactions in any population. On the other hand, it should also be noted that setting the ceiling too low, that is auditing too many large dollar units, can be detrimental to overall efficiency.

---

50. Formula for computing correlation:

Label	Symbol	Formula
Correlation Coefficient (population)	ρ	$$\frac{\sum_{i=1}^{n}\left[(x_i - \bar{X})(y_i - \bar{Y})\right]}{\sqrt{\left[\sum_{i=1}^{n}(y_i - \bar{Y})^2\right] * \left[\sum_{i=1}^{n}(x_i - \bar{X})^2\right]}}$$
Correlation Coefficient (sample)	$\hat{\rho}$	$$\frac{\sum_{i=1}^{n}\left((x_i - \bar{x})(y_i - \bar{y})\right)}{\sqrt{\left[\sum_{i=1}^{n}(y_i - \bar{y})^2\right] * \left[\sum_{i=1}^{n}(x_i - \bar{x})^2\right]}}$$

51. Reference *Sampling Techniques* (Cochran) at pages 132-134, and particularly the comment on page 133: "The results for the regression model indicate that unless ρ exceeds .95, little reduction in variance is to be expected beyond $L = 6$." This means if stratification is on the book values (variable x_i), that correlation (ρ) with the tax errors (y_i) is less than .95, that no reduction can be expected if more than six strata are constructed. In most tax compliance populations, the book values will be correlated with the tax errors in the range where ρ can be expected to be from 0 (no correlation) to .30 (some correlation). Rarely does ρ approach anything greater than .5.

Some agencies also set a ***floor*** amount. Any sampling unit with a book value below the floor will be totally ignored (not sampled or audited at all). In the final evaluation, no projection will be made to units with book values below the floor. It is thought that at some point, that tax (or taxable) error in the low dollar transactions will be insignificant, and even if the error rate in relative terms in that part of the population is large, it will be immaterial to the audit when measured in absolute terms. The use of the floor is not universal to all agencies. Some agencies have strong objections to the use of floors and will not set a floor.[52] Still, assuming that use of the floor can be justified, in some cases, there are problems in how it is applied. It is not unheard of for an agency to set a floor and not sample any units below the floor but then turn around and project into those units below the floor amount. This violates a fundamental concept of statistical sampling. A conclusion or estimate can be extended only to those units that had a possibility being sampled. Since by definition, those units below the floor had no chance at selection, a statistical statement cannot be made about these units. That is, generalizations made from units that had a chance at being sampled cannot be objectively extended to those units that had no chance at being sampled.

In tax compliance audits, the variable that is used to construct strata is the book value. Construction of the strata can be made many using several different methods. The most common method found in practice is described in *Sampling Techniques* (Cochran) at pages 127-131, and is sometimes referred to variously as the ***cumulative frequency method***, or the ***cumulative square root of the frequency***. This method has been found to provide good results, but unfortunately it is rather unwieldy. On the other hand, the process can easily be (in fact has been) automated by computer software packages. Other more straight forward methods, such as proportional weighting and others are in use or have been proposed.[53] As mentioned previously, in a tax compliance audit, stratifying the sampling frame will really only be viable if the auditor has the electronic file of book values. This underscores the importance of the use of computers and advanced sampling methods. Stratification usually provides more efficient samples (and those gains are usually material).

Finally, in a stratified sampling frame, the estimators are much more complex (see appendix). The mean-per-unit and difference estimators have unbiased forms that can be applied. Both the ratio and regression estimators come in two different forms where the point estimates and estimated standard error will not be the same: there are the ***separate*** or ***combined*** forms for these estimators. The separate and combined forms for both the ratio and regression estimators will have some bias. In fact, it is thought that the combined form of ratio and regression will generally exhibit less bias for most

---

52. It is argued that by setting a floor, this is bad tax precedence in that taxpayer's no longer has any incentive to be compliant on any transactions that will likely never be reviewed by the auditor.
53. One interesting method has been proposed in an article: *Contaduría y Administración*, 11 (September-December 2004), Universidad Nacional Autonoma de Mexico (UNAM), "Geometric Stratification of Accounting Data," by Patricia Gunning, Jane Horgan, and William Yancey. The authors of the article claim it does as well as the cumulative frequency method. Others who have looked closely at this method dispute this (including this author).

tax compliance populations.[54] Unfortunately, the combined form is more complex, and the overall point estimate will not be the sum of the individual point estimates coming from the individual strata.[55]

§2.10 PROBLEMS IN APPLYING STATISTICAL SAMPLING

Not all audits require the auditor to sample. Further, not all audits where sampling is needed are conducive to statistical sampling. Occasionally, an auditor in a tax compliance audit will apply an overly burdensome sampling procedure, when a more simple procedure would have sufficed. But by far, the biggest problem in tax compliance auditing is that the auditor uses a judgmental sample, when a statistical could have—and should have—been done. This is due to a variety of reasons. In some agencies, while some statistical sampling is performed, there is no monitoring of sampling decisions made by auditors. Where this is true, statistical sampling techniques tend to be underutilized. Agency management is often unconcerned by such matters and leave these decisions to the field auditor. Since statistical sampling requires more work up front—and there is no consequence for applying a less appropriate sampling procedure, many auditors are reluctant to do this work (there is a desire to immediately dive into the documentation, rather than planning a better sample). Older auditors especially (many in lead positions), who have used judgmental sampling throughout their careers, believe incorrectly that a judgmental sample to always be just as good—"so why bother?" Even in agencies that perform statistical sampling, many auditors do not receive the training or necessary software to perform a statistical sample. Finally, it must be said, that some auditors appear incapable of learning how to do statistical sampling. Some may lack the ability to understand—or at least show a complete disinterest in learning the theory. While others do not have the perquisite computer skills or are just intimidated by large data sets.

Statistical sampling requires an investment by the agency and the auditor. There must be considerable thought to sampling policy by the agency (see the discussion in the next section). The agency must supply the necessary training and software to the auditor to perform these tasks. The auditor needs to understand the policy, software, and be willing to learn new procedures to perform tasks previously done by less sophisticated procedures. Some agencies, particularly in smaller states, do not wish to tackle these problems.

The type of populations found in tax compliance auditing also present problems. They tend to be highly variable, requiring large samples and sophisticated sampling designs to give adequate or useful results. The above discussion on low error rate populations applies in most every audit. Usually, statistical sampling is more difficult for the tax compliance auditor to execute, when compared to other areas where these

---

54. The IRS, Multistate Tax Commission, and some state agencies use the combined ratio and combined regression estimator instead of the separate ratio and separate regression estimators for this reason.
55. For mean-per unit, difference, separate ratio and separate regression, the overall point estimate will be the sum of the point estimates coming from the individual strata.

techniques are applied. This and the lack of buy-in by agency managers probably accounts for the slow adoption, even though the computer technology has been around for some time.

One problem that is always difficult to deal with is expectations of taxpayers and auditors. When statistical sampling results are expressed in other applications, the "error bars" are often stated in percentage terms of a few percent—"plus or minus five percent." In fact, such narrow confidence intervals are difficult to achieve in tax compliance applications. Oftentimes auditors who come to appreciate the lack of precision inherent to tax compliance populations (that they were otherwise previously oblivious to) will often do less sampling as a response, not statistical sampling. Another expectation is that best method is the one that is simple and easy to follow. In statistical sampling, this is not so, and so some auditors automatically presume that statistical sampling is not the optimal method when, in fact, it often is. Indeed, in some agencies, there is often a push by well-meaning bureaucrats to produce government reports in simple language, that is to make the audit reports so the "average person" can understand them. Obviously, the language of statistics is not elementary, and use of statistical sampling is not compatible with such efforts.

Other problems have been already been described. Some auditors will take a probability sample yet not perform a statistical evaluation and call the procedure a "statistical sample." Still others will, to make an effort to comply with the evaluation requirement, compute a confidence interval on the book values, not the estimate upon which the basis of the assessment. Sometimes, the confidence interval is computed on the estimated value as is required, but then not disclosed to the taxpayer because its relative width. If sampling policies are silent or not enforced, documentation can be lacking or not provided to the taxpayer.

APPENDIX I: EXPLANATION OF BOOK VALUE, ERROR VALUE, & AUDITED VALUE AND RESPECTIVE SAMPLE VALUES FOR EXAMPLE

The example in this section uses a specific sample (which is entirely made up). The following table provides a complete listing of the audited sample results. Note that since the taxpayer is mainly in compliance, relatively few error values (y_i) have a value other than zero. In the table, we can see that the book value (x_i) is the sum of the error value and the audited value. Also, the "audited value," defined in the text as the difference between the book value and error value, usually has no greater meaning in tax compliance. However, within the financial audit arena, the audited value represents what the book value should be according to the auditor's opinion. A strong word of caution here must be made with respect to the audited value. In tax compliance auditing, this is nearly never the case. The audited value in almost always merely a plug value between some book value and the actual tax error. Therefore, in nearly all tax compliance applications, the auditor will only record the book value (x_i) and error value (y_i) in the audit papers. The computer will generate the audit values for purposes of statistical evaluation, particularly for the difference estimator. Since the difference estimator virtually never performs well in low error populations and is almost never used, the audit values of the sample will have no bearing on final sample projection used to make an assessment.

The book value (x_i) in tax compliance is sometimes referred to as the **recorded value**, **reported value**, or **examined amount**. It could represent a deduction on the return, a book entry in an account, or just a line item in a listing of sales or purchases. A popular term in sales and use tax auditing for book value is **invoice amount**. Many sampling frames in sales and use tax audits are listings of sales or purchases.

The error value is the actual tax error associated with the chosen book value. If the auditor is unable to determine what error value is associated with any book value, another book value should be selected when constructing the sampling frame. If no such book value exists, tax difference auditing cannot be performed. Usually in that event, the auditor must then project total taxable amount via statistical formula similar to that found herein, than subtract out what was reported to arrive at the total tax difference. In practice, projection of total taxable is rarely done, and tax difference auditing (establishing a tax error, or y_i for each sampling unit in the sample) is the norm.

Sample Item (i)	Book Value (x_i)	Error Value (y_i)	Audited Value (x_i-y_i)	Sample Item (i)	Book Value (x_i)	Error Value (y_i)	Audited Value (x_i-y_i)
1	1,399	0	1,399	51	643	643	0
2	300	0	300	52	1,300	0	1,300
3	1,389	0	1,389	53	208	0	208
4	1,028	0	1,028	54	480	0	480
5	1,228	0	1,228	55	634	0	634
6	1,025	0	1,025	56	322	0	322
7	550	0	550	57	1,732	0	1,732
8	217	0	217	58	1,322	0	1,322
9	314	0	314	59	1,609	0	1,609
10	461	0	461	60	2,000	0	2,000
11	278	0	278	61	1,050	0	1,050
12	490	0	490	62	1,129	0	1,129
13	135	0	135	63	765	0	765
14	1,559	1,559	0	64	339	0	339
15	127	0	127	65	1,625	0	1,625
16	1,336	0	1,336	66	351	0	351
17	1,650	0	1,650	67	1,648	0	1,648
18	558	0	558	68	1,916	0	1,916
19	361	0	361	69	51	0	51
20	350	350	0	70	1,431	0	1,431
21	1,980	0	1,980	71	1,758	0	1,758

Chapter 2: Statistical Sampling in Tax Compliance Auditing in the USA Appendix II

Sample Item ($_i$)	Book Value (x_i)	Error Value (y_i)	Audited Value (x_i-y_i)	Sample Item ($_i$)	Book Value (x_i)	Error Value (y_i)	Audited Value (x_i-y_i)
22	503	0	503	72	1,412	0	1,412
23	150	0	150	73	280	0	280
24	1,280	0	1,280	74	137	0	137
25	1,691	0	1,691	75	1,361	1,361	0
26	442	0	442	76	650	0	650
27	1,882	0	1,882	77	433	433	0
28	1,752	0	1,752	78	403	0	403
29	1,050	0	1,050	79	570	0	570
30	273	0	273	80	113	0	113
31	212	0	212	81	1,645	0	1,645
32	425	0	425	82	1,803	0	1,803
33	1,235	0	1,235	83	620	0	620
34	511	0	511	84	1,450	1,450	0
35	320	0	320	85	167	167	0
36	900	0	900	86	738	0	738
37	1,649	0	1,649	87	1,456	0	1,456
38	364	0	364	88	275	0	275
39	127	0	127	89	675	0	675
40	1,971	0	1,971	90	660	0	660
41	1,080	0	1,080	91	299	0	299
42	1,391	0	1,391	92	668	0	668
43	81	0	81	93	1,634	0	1,634
44	366	0	366	94	419	0	419
45	1,874	0	1,874	95	1,552	0	1,552
46	1,346	0	1,346	96	610	0	610
47	500	0	500	97	1,760	0	1,760
48	63	0	63	98	417	0	417
49	605	0	605	99	379	0	379
50	1,995	1,995	0	100	1,001	0	1,001

APPENDIX II: FORMULA FOR STRATIFIED RANDOM SAMPLING

The following formula describe what is found most often in practice, statistical evaluations based on a stratified sampling frame. These formula are mainly derived from *Sampling Techniques* (Cochran). The notation style below is compatible with Cochran's style. Financial auditors also use these formulas as well when performing classical variables sampling. The formula can be found in Appendix 2 and 7 of

Appendix II — Robert Schauer

Statistical Auditing (Roberts). The notation style found in Robert's book is not entirely compatible with the notation below. Note that in Roberts book, book values (referred to as recorded amounts) are described by y_i (individual book values) and Y (total book value), errors (referred to as differences) as d_i (individual error values) and D (total error value), and audited amounts as x_i (individual audited values) and X (total book value).

Label	Symbol	Formula
General formula:		
Number of Strata Sampled	L	
Stratum Count	N_h	[count of sample units in the population for the hth stratum]
Sample Count	n_h	[count of sample units in the sample of the hth stratum]
Sampling Fraction in the hth Stratum	f_h	n_h / N_h
Finite Population Correction in the hth Stratum	fpc_h	$\sqrt{1 - f_h}$
Book Value	$x_{h,i}$	[the ith book value in the hth stratum]
Total Book Value in the hth Stratum	X_h	$\sum_{i=1}^{N_h} x_{h,i}$
Total Book Value in the Sampled Strata	X_{st}	$\sum_{h=1}^{L} X_h$
Mean Book Value of the Sample in the hth Stratum	$\bar{x}_h$	$\dfrac{\sum_{i=1}^{n_h} x_{h,i}}{n_h}$
Variance of the Book Values in the hth Stratum	$s^2_{x_h}$	$\dfrac{\sum_{i=1}^{n_h} (x_{h,i} - \bar{x}_h)^2}{n_h - 1}$
Taxable Error Value	$y_{h,i}$	[the ith taxable error value in sample of the hth stratum]

Chapter 2: Statistical Sampling in Tax Compliance Auditing in the USA Appendix II

Label	Symbol	Formula
Total Taxable Error Value in the hth Stratum	Y_h	$\sum_{i=1}^{N_h} y_{h,i}$
Total Taxable Error Value in the Sampled Strata	Y_{st}	$\sum_{h=1}^{L} Y_h$
Mean Taxable Error Value of the Sample in the hth Stratum	$\bar{y}_h$	$\dfrac{\sum_{i=1}^{n_h} y_{h,i}}{n_h}$
Variance of the Taxable Error Values in the hth Stratum	$s^2_{y_h}$	$\dfrac{\sum_{i=1}^{n_h}(y_{h,i}-\bar{y}_h)^2}{n_h-1}$
Audited Value		$x_{h,i} - y_{h,i}$
Total Audited Value in the hth Stratum		$X_h - Y_h$
Total Audited Value in the Sampled Strata		$\sum_{h=1}^{L}(X_h - Y_h)$
Mean Audited Value of the Sample in the hth Stratum		$\bar{x}_h - \bar{y}_h$
Variance of the Audited Values in the hth Stratum		$\dfrac{\sum_{i=1}^{n_h}\left[(x_{h,i}-y_{h,i})-(\bar{x}_h-\bar{y}_h)\right]^2}{n_h-1}$
Stratified Mean-per-unit Estimation of Total Book Value:		
Estimated Total Book Value in the hth Stratum	$\hat{X}_h$	$N_h \bar{x}_h$
Estimated Total Book Value for all Sampled Strata	$\hat{X}_{st}$	$\sum_{h=1}^{L} \hat{X}_h$

Label	Symbol	Formula
Standard Error of the hth Stratum (book values)	$s_{\hat{X}_h}$	$\dfrac{N_h s_{x_h} fpc_h}{\sqrt{n_h}}$
Overall Standard Error of the Estimated Total Book Value	$s_{\hat{X}_{st}}$	$\sqrt{\sum_{h=1}^{L} s_{\hat{X}_h}^2}$
Effective Degrees of Freedom	n_e	$\dfrac{s_{\hat{X}_{st}}^4}{\sum_{h=1}^{L}\left[s_{\hat{X}_h}^4/(n_h-1)\right]}$ or $\dfrac{\left(\sum_{h=1}^{L} g_h s_{x_h}^2\right)^2}{\sum_{h=1}^{L} \dfrac{g_h^2 s_{x_h}^4}{n_h-1}}$
Student's t (using effective degrees of freedom)	t_{n_e}	[normal deviate]
Confidence Interval Estimated Total Book Value		$\hat{X}_{st} \pm \left(t_{n_e} * s_{\hat{X}_{st}}\right)$
Stratified Mean-per-unit Estimation of Total Taxable Error Value:		
Estimated Total Taxable Error in the hth Stratum	$\hat{Y}_{h,m}$	$N_h \bar{y}_h$
Estimated Total Taxable Error for all Sampled Strata	$\hat{Y}_{st,m}$	$\sum_{h=1}^{L} \hat{Y}_{h,m}$
Standard Error of the hth Stratum	$s_{\hat{Y}_h,m}$	$\dfrac{N_h s_{y_h} fpc_h}{\sqrt{n_h}}$
Overall Standard Error of the Estimated Total Taxable Error for all Sampled Strata	$s_{\hat{Y}_{st},m}$	$\sqrt{\sum_{h=1}^{L} s_{\hat{Y}_h,m}^2}$

Chapter 2: Statistical Sampling in Tax Compliance Auditing in the USA Appendix II

Label	Symbol	Formula
Effective Degrees of Freedom	n_e	$\dfrac{s_{\hat{Y}_{st,m}}^4}{\sum_{h=1}^{L}\left[s_{\hat{Y}_h,m}^4/(n_h-1)\right]}$ or $\dfrac{\left(\sum_{h=1}^{L} g_h s_{y_h}^2\right)^2}{\sum_{h=1}^{L}\dfrac{g_h^2 s_{y_h}^4}{n_h-1}}$ where $g_h = \dfrac{N_h(N_h-n_h)}{n_h}$
Confidence Interval Estimated Total Taxable Error		$\hat{Y}_{st,m} \pm \left(t_{n_e} s_{\hat{Y}_m,d}\right)$
Stratified Difference Estimation of Total Taxable Error Value:		
Estimated Total Taxable Error in the *h*th Stratum	$\hat{Y}_{h,d}$	$X_h - N_h(\bar{x}_h - \bar{y}_h)$
Estimated Total Taxable Error for all Sampled Strata	$\hat{Y}_{st,d}$	$\sum_{h=1}^{L} \hat{Y}_{h,d}$
Standard Deviation of the *h*th Stratum (of audited values)	$s_{(x-y)_h,d}$	$\sqrt{\dfrac{\sum_{i=1}^{n_h}\left[(x_{h,i}-y_{h,i})-(\bar{x}_h-\bar{y}_h)\right]^2}{n_h-1}}$
Standard Error of the *h*th Stratum	$s_{\hat{Y}_h,d}$	$\dfrac{N_h s_{(x-y)_h,d} \, fpc_h}{\sqrt{n_h}}$
Overall Standard Error of the Estimated Total Taxable Error for all Sampled Strata	$s_{\hat{Y}_{st},d}$	$\sqrt{\sum_{h=1}^{L} s_{\hat{Y}_h,d}^2}$
Effective Degrees of Freedom	n_e	$\dfrac{s_{\hat{Y},d_{st}}^4}{\sum_{h=1}^{L}\left[s_{\hat{Y}_h,d_h}^4/(n_h-1)\right]}$
Confidence Interval Estimated Total Taxable Error		$\hat{Y}_{st,d} \pm \left(t_{n_e} * s_{\hat{Y}_{st},d}\right)$
Stratified Separate Ratio Estimation of Total Taxable Error Value:		

Appendix II

Label	Symbol	Formula
Estimate of the Ratio of the Taxable Error in the hth Stratum	$\hat{R}_{S_h}$	$\bar{y}_h / \bar{x}_h$
Estimated Total Taxable Error in the hth Stratum	$\hat{Y}_{Rs_h}$	$X_h \hat{R}_{S_h}$
Estimated Total Taxable Error for all Sampled Strata	$\hat{Y}_{Rs}$	$\sum_{h=1}^{L} \hat{Y}_{Rs_h}$
Standard Deviation	$s_{y_{Rs},h}$	$\sqrt{\dfrac{\sum_{i=1}^{n_h}\left[y_{h,i} - \left(\hat{R}_{S_h} x_{h,i}\right)\right]^2}{n_h - 1}}$
Standard Error of the hth Stratum	$s_{\hat{Y}_{Rs},h}$	$\dfrac{N_h s_{y_{Rs},h} fpc_h}{\sqrt{n_h}}$
Overall Standard Error of the Estimated Total Taxable Error for all Sampled Strata	$s_{\hat{Y}_{Rs}}$	$\sqrt{\sum_{h=1}^{L} s_{\hat{Y}_{Rs},h}^2}$
Confidence Interval Estimated Total Taxable Error		$\hat{Y}_{Rs} \pm \left(z * s_{\hat{Y}_{Rs},h}\right)$ (some jurisdictions compute t_{n_e} for this estimator, and use that instead of z)
Stratified Combined Ratio Estimation of Total Taxable Error Value:		
Estimated Combined Ratio of the Taxable Error for all Sampled Strata	$\hat{R}_C$	$\dfrac{\sum_{h=1}^{L}(N_h \bar{y}_h)}{\sum_{h=1}^{L}(N_h \bar{x}_h)}$
Estimated Total Taxable Error for all Sampled Strata	$\hat{Y}_{Rc}$	$X * \hat{R}_C$
Standard Deviation of the hth Stratum	$s_{y_{Rc},h}$	$\sqrt{\dfrac{\sum_{i=1}^{n_h}\left[(y_{h,i} - \bar{y}_i) - \hat{R}_C(x_{h,i} - \bar{x}_i)\right]^2}{n_h - 1}}$

Chapter 2: Statistical Sampling in Tax Compliance Auditing in the USA Appendix II

Label	Symbol	Formula
Standard Error of the hth Stratum	$s_{\hat{Y}_{Rc,h}}$	$\dfrac{N_h s_{y_{Rc,h}} fpc_h}{\sqrt{n_h}}$
Overall Standard Error of the Estimated Total Taxable Error for all Sampled Strata	$s_{\hat{Y}_{Rc}}$	$\sqrt{\sum_{h=1}^{L} s_{\hat{Y}_{Rc,h}}^2}$
Confidence Interval Estimated Total Taxable Error Value		$\hat{Y}_{Rc} \pm \left(z * s_{\hat{Y}_{Rc}}\right)$ (some jurisdictions compute t_{n_e} for this estimator, and use that instead of z)
Stratified Separate Regression Estimation of Total Taxable Error Value:		
Regression Coefficient of the Taxable Error in the hth Stratum	b_{Gs_h}	$\dfrac{\sum_{i=1}^{n_h}\left[(x_{h,i}-\bar{x}_h)(y_{h,i}-\bar{y}_h)\right]}{\sum_{i=1}^{n_h}(x_{h,i}-\bar{x}_h)^2}$
Estimated Total Taxable Error in the hth Stratum	$\hat{Y}_{Gs_h}$	$N_h \bar{y}_h + b_{Gs_h}\left[X_h - (N_h \bar{x}_h)\right]$
Estimated Total Taxable Error for all Sampled Strata	$\hat{Y}_{Gs}$	$\sum_{h=1}^{L} \hat{Y}_{Gs_h}$
Standard Deviation of the hth Stratum	$s_{y_{Gs,h}}$	$\sqrt{\dfrac{\sum_{i=1}^{n_h}\left[(y_{h,i}-\bar{y}_h)-b_{Gs_h}(x_{h,i}-\bar{x}_h)\right]^2}{n_h-2}}$
Standard Error of the hth Stratum	$s_{\hat{Y}_{Gs,h}}$	$\dfrac{N_h s_{y_{Gs,h}} fpc_h}{\sqrt{n_h}}$
Overall Standard Error of the Estimated Total Taxable Error for all Sampled Strata	$\hat{Y}_{Gs}$	$\sqrt{\sum_{h=1}^{L} s_{\hat{Y}_{Gs,h}}^2}$

Label	Symbol	Formula
Confidence Interval Estimated Total Taxable Error		$\hat{Y}_{Gs} \pm \left(z * s_{\hat{Y}_{Gs,h}}\right)$ (some jurisdictions compute t_{n_e} for this estimator, and use that instead of z)
Stratified Combined Regression Estimation of Total Taxable Error Value:		
Covariance of the Book Values and the Taxable Error in the hth Stratum	$Cov(x,y)_h$	$\dfrac{\sum_{i=1}^{n_h}\left[(x_{h,i}-\bar{x}_h)(y_{h,i}-\bar{y}_h)\right]}{n_h - 1}$
Combined Regression Coefficient	b_{Gc}	$\dfrac{\sum_{h=1}^{L}\left[N_h(N_h - n_h)\dfrac{Cov(x,y)_h}{n_h}\right]}{\sum_{h=1}^{L}\left[N_h(N_h - n_h)\dfrac{s_{x_h}^2}{n_h}\right]}$
Estimated Total Taxable Error for all Sampled Strata	$\hat{Y}_{Gc}$	$\sum_{h=1}^{L} N_h \bar{y}_h + b_{Gc}\left(X - \sum_{h=1}^{L} N_h \bar{x}_h\right)$
Standard Error of the hth Stratum	$s_{y_{Gc,h}}$	$\sqrt{s_{y_h}^2 - 2b_{Gc}Cov(x,y)_h + b_{Gc}^2 s_{x_h}^2}$
Standard Error of the hth Stratum	$s_{\hat{Y}_{Gc,h}}$	$\dfrac{N_h s_{y_{Gc,h}} fpc_h}{\sqrt{n_h}}$
Overall Standard Error of the Estimated Total Taxable Error for all Sampled Strata	$s_{\hat{Y}_{Gc}}$	$\sqrt{\sum_{h=1}^{L} s_{\hat{Y}_{Gc,h}}^2}$
Confidence Interval Estimated Total Taxable Error		$\hat{Y}_{Gc} \pm \left(z * s_{\hat{Y}_{Gc}}\right)$ (some jurisdictions compute t_{n_e} for this estimator, and use that instead of z)

§2.11 REFERENCES

Abdel-khalik, A. Rashad & Ira Solomon (eds.), *1984 Auditing Research Symposium*, Office of Accounting Research, University of Illinois at Urbana/Champaign (1985).

AICPA SAS 39 - *Audit Guide.*

Auditing Practice Release; *Audit Sampling*, AICPA (1999).

Arens, Alvin & James K Loebbecke, *Applications of Statistical Sampling to Auditing*, Prentice Hall, Englewood CA (1981).

Arens, Alvin & James K Loebbecke, *Auditing: An Integrated Approach*, Prentice Hall, Englewood CA (1994).

Arkin, Herbert, *Sampling Methods for the Auditor; an Advanced Treatment* (1982).

Arkin, Herbert *Handbook of Sampling for Auditing and Accounting*, McGraw Hill, New York, 1974.

California State Board of Equalization , *Audit Manual, Chapter 13, Statistical Sampling; Sales and Use Tax* (January 2000), www.boe.ca.gov/pdf/fam-13.pdf.

Cochran, William, *Sampling Techniques*, John Wiley and Sons (1977).

Cummings, Rocky, *Too Much Information*, Tax Trends, CCH (2007).

Deming, William Edwards, *Some Theory of Sampling*, Dover Publications, Inc (1950).

Deming, W. Edwards, "Standards of Probability Sampling for Legal Evidence," *The American Statistician* (February 1958).

Department of Health & Human Services, OIG—Office of Audit Services *RAT-STATS Companion Manual* (September 2001), oig.hhs.gov/organization/oas/ratstats/CompManual2007.pdf.

Department of Health and Human Services, Office of Inspector General, Office of Audit Services, *RAT-STATS 2007 User Guide, version 2* (Revised October 2004), oig.hhs.gov/organization/oas/ratstats/UserGuide2007.pdf

Efron, Bradley & Robert J. Tibshirani, *An Introduction to the Bootstrap*, Chapman & Hall/CRC, 1993

Federation of Tax Administrators, *Sampling for Sales and Use Tax Compliance* (December 2002), http://www.taxadmin.org/fta/pub/sample.pdf.

Federation of Tax Administrators Task Force on EDI Audit and Legal Issues for Tax Administration, *Appendix A: Survey of State Sampling Practices* (January 2004), www.taxadmin.org/fta/pub/Samp2004.pdf.

Florida Department of Revenue, *Auditing in an Electronic Environment (e-auditing) and Stratified Statistical Sampling* (2002), booklet #GT-300034, dor.myflorida.com/dor/forms/2002/gt300034.pdf.

Gunning, Patricia, Jane Horgan & William Yancey. "Geometric Stratification of Accounting Data," *Contaduría y Administración*, 11 (September-December 2004), Universidad Nacional Autonoma de Mexico (UNAM).

Guy, Dan M., *An Introduction to Statistical Sampling in Auditing*, John Wiley and Sons, 1981.

Hansen, Morris H., William N. Hurwitz & William G. Madow, *Sample Survey Methods and Theory, Volumes #1 and #2*, John Wiley & Sons (1993).

Hitzig, Neal B., "Statistical Sampling Revisited," *The CPA Journal* (2004).

International Federation of Accountants (IFAC), International Standards on Accounting (ISA) 530.

IRS Training Publication: *Advanced Statistical Sampling*, Training 3174-002 (Revised May 1992) TPDS 87030A.

IRS Training Publication: *Basic Statistical Sampling*, Training 3172-001 (Revised August 1993) TPDS 871251.

Jennings, Harold & Robert Schauer, *Why Use Statistical Sampling*, MTC Review (Summer 2008), www.mtc.gov.

Kish, Leslie, *Survey Sampling*, John Wiley & Sons (1995).

Kvanli, Alan & Robert Schauer, "The Bootstrap, What the Government Auditor Should Know," *Journal of Government and Financial Management* (Fall 2002) .

Kvanli, Alan & Robert Schauer "Is Your Agency Too Conservative? Deriving More Reliable Confidence Intervals," *Journal of Government and Financial Management* (Summer 2005).

Leslie, Donald A., Albert D. Teitlebaum & Rodney J. Anderson, *Dollar-unit Sampling, A practical Guide for Auditors*, CCH (1979).

Levy, Paul & Stanley Lemeshow, *Sampling of Populations: Methods and Applications*, 3rd Edition, John Wiley & Sons, New York (1999).

Lohr, Sharon, *Sampling: Design and Analysis*, Duxbury Press (1999).

McRae, T.W., *Statistical Sampling for Audit and Control*, John Wiley & Sons (1974).

Multistate Tax Commission, *Sampling Policy & Guideline Manual* (July 2008), www.mtc.gov/Audit.aspx?id=612

New York State Department of Taxation & Finance New York State Department of Taxation & Finance, *Computer Assisted Audits, Guidelines and Procedures for Sales Tax Audits* (October 2001), Publication 132, www.tax.state.ny.us/pdf/publications/sales/pub132_1001.pdf.

Newman, Maurice S., *Accounting Estimates by Computer Sampling*, John Wiley & Sons, New York (1982).

Olken, Frank, *Random Sampling from Databases*, University of California at Berkley (1993).

Owen, Art B., *Empirical Likelihood*, Chapman & Hall/CRC, 2001.

Roberts, Donald A., *Statistical Auditing*, AICPA (1978).

Scheaffer, Richard, William Mendenhall & R. Lyman Ott, *Elementary Survey Sampling*, 5th edition, Duxbury Press.

Tennessee Department of Revenue, Audit Division, *Statistical Sampling for Sales and Use Tax Audits*, Revised (April 1999), www.state.tn.us/revenue/tntaxes/sales/statsamplingapr06.pdf.

Washington State Department of Revenue, *Statistical Sampling Manual* (revised January 2008).

Yancey, Will, *Statistical Sampling in Sales and Use Tax Audits*, CCH (2002).

CHAPTER 3
Legal Issues Surrounding Sampling

Robert F. van Brederode

§3.01 INTRODUCTION

Under the tax law of most countries, the starting point of a tax audit is the administration of the taxpayer, including the tax returns filed. Before resorting to external indices to estimate tax, following the example of New York case law for sales tax, the taxing authority must first request[1] and thoroughly examine[2] the taxpayer's books and records for the entire period of the proposed assessment.[3] The purpose of the examination is to determine, through verification drawn independently from within these records[4] that they are, in fact, so insufficient that it is virtually impossible to verify taxable sales receipts and conduct a complete audit[5] from which the exact amount of tax due can be determined.[6] Where the taxing authority follows this procedure, thereby demonstrating that the records are incomplete or inaccurate, it may resort to external indices to estimate tax.[7] The estimate methodology utilized must be reasonably calculated to reflect taxes due,[8] but exactness in the outcome of the audit method is not required.[9] The taxpayer bears the burden of proving with clear and

---

1. *Christ Cella, Inc. v. State Tax Commn.*, 102 AD2d 352, 477 NYS2d 858.
2. *King Crab Rest. v. Chu*, 134 AD2d 51, 522 NYS2d 978.
3. *Adamides v. Chu*, 134 AD2d 776, 521 NYS2d 826, lv denied 71 NY2d 806, 530 NYS2d 109.
4. *Giordano v. State Tax Commn.*, 145 AD2d 726, 535 NYS2d 255; *Urban Liqs. v. State Tax Commn.*, 90 AD2d 576, 456 NYS2d 138; *Meyer v. State Tax Commn.*, 61 AD2d 223, 402 NYS2d 74, lv denied 44 NY2d 645, 406 NYS2d 1025; and *Hennekens v. State Tax Commn.*, 114 AD2d 599, 494 NYS2d 208.
5. *Chartair, Inc. v. State Tax Commn.*, 65 AD2d 44, 411 NYS2d 41, 43; *Christ Cella, Inc. v. State Tax Commn.*, supra, note 1.
6. *Mohawk Airlines v. Tully*, 75 AD2d 249, 429 NYS2d 759, 760.
7. *Urban Liqs. v. State Tax Commn.*, supra, note 4.
8. *W.T. Grant Co. v. Joseph*, 2 NY2d 196, 159 NYS2d 150, cert denied 355 US 869, 2 L Ed 2d 75.
9. *Markowitz v. State Tax Commn.*, 54 AD2d 1023, 388 NYS2d 176, affd 44 NY2d 684, 405 NYS2d 454; Cinelli, Tax Appeals Tribunal, September 14, 1989.

convincing evidence that the assessment is erroneous[10] or that the audit methodology is unreasonable.[11] In addition, considerable latitude is given an auditor's method of estimating sales under such circumstances as exist in each case.[12] The burden then rests upon the taxpayer to demonstrate that the method of audit or the amount of the assessment was erroneous.[13]

The standard procedure for sales tax in New York State has a stated preference for thorough examination. Estimation appears at the horizon once the books and records have been verified as inaccurate and, therefore, as unreliable for determining the tax position of the taxpayer under audit. Sampling, of course, is essentially a method of estimation, but one that requires accurate books and records. In fact, there is a direct correlation between the level of accuracy of the sample and the accuracy level of the underlying documents. That does not mean that the New York approach excludes the use of sampling. The policy described afore is limited to the use of external indices. Estimation on the basis of external indices is only allowed when the books and records are inaccurate. However, sampling is an estimation method based on *internal* indices. So the starting question is whether sampling should be allowed (section § 3.02) and we approach this issue from different perspectives. First we examine the extent to which the allocation of the burden of proof should play a role in answering this question (section § 3.02[A]). In other words, does it make a difference for the permissibility of sampling whether the burden of proof rests with the taxpayer or the taxing authority? Second, we address the issue of whether the use of sampling should be depending on the type of tax, using a number of Netherlands Supreme Court rulings to frame the discussion (section § 3.02[B]). Next, we discuss the influence that the confidence interval should have in determining the point of assessment (section § 3.02[C]).

The result of sampling is not the exact determination of the under-or overpayment by a taxpayer but only an estimate of her position. There is a risk of inaccuracy which needs to be measured and found to be acceptable. Accuracy can be improved by using larger samples, but this is in most cases impractical and may undermine the purpose of sampling itself when the size of the sampling approaches that of a full audit. In our opinion, the permissibility of sampling is closely related to where the amount of ultimate assessment is located on the line connecting the opposite sides of the interval width.

In section § 3.02[D] we discuss the phenomenon of sample agreements as the basis of setting the terms under which sampling audits can be conducted. Last but not least, we will take a closer look at the permissibility of sampling from the perspective of fair play (section § 3.02[E]).

An urgent legal question is further whether the results of a sample test can be applied outside its direct scope, i.e., whether extrapolation should be allowed to earlier years not covered by the audit (section § 3.03). Another important legal aspect is

---

10. *Scarpulla v. State Tax Commn.*, 120 AD2d 842, 502 NYS2d 113.
11. *Surface Line Operators Fraternal Org. v. Tully*, 85 AD2d 858, 446 NYS2d 451; Cousins Serv. Station, Tax Appeals Tribunal, August 11, 1988.
12. *Grecian Sq. v. New York State Tax Commn.*, 119 AD2d 948, 501 NYS2d 219, 221.
13. Your Own Choice, Inc., Tax Appeals Tribunal, February 20, 2003.

whether and, if so, how the results from sampling can be applied to determine interest payments and penalties (section § 3.04). The last section contains a summary of our findings and conclusions.

§3.02 PERMISSIBILITY OF SAMPLING

[A] Burden of Proof

Before a tax authority decides to use statistical sampling in tax audits, it needs to determine whether such an audit method is legally allowed under the law and precedence. Or, from the flip side of the coin, a taxpayer needs to determine whether she can oppose the principle of using a sample to estimate a tax liability in a government audit. That is an open question probably subject to further litigation in most jurisdictions. Fortunately, there exists some case law in the US and in the Netherlands— the forerunner in Europe in the application of sampling in tax auditing—to provide us as to this question with the insights of the judicial branch of government. The acceptability of sampling as an audit method seems to have a prima facie correlation with the burden of proof. If a taxpayer's books and records are acceptable, then the presumption should be that her tax returns are acceptable as well, unless proven otherwise. However, that should not necessarily mean that the burden of proof always and completely rests with the taxing authority. Rules on the burden of proof in tax vary among jurisdictions, but it seems reasonable to assume that all jurisdictions try to establish a reasonable balance when dividing the burden of proof between the taxing authority and the taxpayer. For example, it seems reasonable that the burden of proof would rest with the taxing authority if it takes the position that a taxpayer's profit or income is higher than reported, but that a taxpayer needs to proof her eligibility to and the calculative correctness of a deduction or exemption claimed. It appears reasonable for a taxing authority to resort to sampling for the purpose of estimating the size of an alleged mistake for which the proof of burden rests with the taxpayer, because the latter can counter any assessment of error by conducting a detailed audit of her books and records. It is fair to ask of the taxpayer to undertake this, because the burden of proof is hers. However, if the burden of proof rests with the taxing authority, the question arises whether an estimation of the error is acceptable over exact calculation. Indeed, the taxpayer can still conduct a detailed audit of her own books and records to counter prove the estimation, but that would in actuality lead to a *contra legem* shifting of the burden of proof onto the taxpayer. There is no justification for that. Nevertheless, in our opinion, an estimation of error can be accepted in case the taxing authority settles for the minimum error position based on the width of the confidence interval (taking the 1-sided confidence level). On the other hand, if the burden of proof rest with the taxpayer, a tax assessment calculation based on the mid-range of the confidence interval (point estimate) appears to provide a fair balance between the interests of the state and the taxpayer, given that the latter retains the right to make exact calculations on the basis of a detailed audit of books and records.

[B] Type of Tax

The question may arise whether the permissibility of sampling audits should depend on the type of tax under review. The Netherlands Supreme Court had to resolve this question in three cases covering wage tax, value added tax (VAT), and passenger car tax.[14]

The passenger car tax is basically an excise type of tax levied on the person in whose name the car or motorcycle is or will be registered.[15] However, if the registration is filed by someone else than the person in whose name the vehicle will be registered, the registrant is required to file the tax return and pay the tax on behalf of that person.[16] In practice, it is the importer of the cars and motorcycles who will register the vehicles. Many buyers of cars require the dealer to customize the car with accessories and gadgets unknown to the importer at the time of registration.[17] As a result, the importer has underpaid passenger car tax since these accessories are also subject to the tax. In principle, the importer can be assessed for the underpaid tax, but she has legal regress on the person in whose name the car has been registered. Because the importer is paying a tax for which another person is legally liable, the Supreme Court did not accept statistical sampling. The importer can only be required to pay the exact amount due by each individual person in whose name the car is registered. Statistical sampling does not provide individualized information and, therefore, does not allow the importer to exercise her right of regress. Essential in this case is that the importer has no means to determine the additional tax liabilities for each individual car owner because her own administration does not include information on the dealer provided accessories.

The Netherlands Supreme Court came to a different conclusion for VAT and wage tax.

The wage or payroll tax is closely related to the individual income tax and is, in essence, a pretax of the income tax. Individuals report their individual incomes at the end of a calendar year and pay the tax through filing a return. For most individuals, wages and salaries are the main source of income and by mandating withholding of tax by the employer from these wages or salaries states can raise the level of assurance that (sufficient) tax will, indeed, be paid. Most countries apply this system[18] that also allows the state to collect the tax earlier than through the annual return filing procedure.[19]

---

14. *Hoge Raad*, 14 March 2008, nos. 39.866, *NFTR*, 2008/549; 40.474, *NFTR*, 2008/548; and 40.806, *NFTR*, 2008/547.
15. Article 5(1), Wet Belasting van personenauto's en rijwielen (Act on passenger cars and motorcycles tax).
16. Article 7, Wet Belasting van personenauto's en rijwielen (Act on passenger cars and motorcycles tax).
17. This problem has been repaired by a change of law, see Article 9(9) Wet van personenauto's en rijwielen (Act on passenger cars and motorcycles tax).
18. In the US, mandatory withholding was established through the Current Tax Payment Act of 1943, invented by Milton Friedman who was employed by the Treasury department at the time and still a Keynesian.
19. And to tax more, because installment payment reduces the taxpayers' consciousness of the total tax burden. This in combination with the psychologically anesthetizing effect of parting from money that one has not effectively received weakens their resistance as well.

When filing his annual income tax return, the taxpayer deducts the wage tax earlier withheld by his employer and remits the difference to the taxing agency or receives a refund in case of overpayment. The Netherlands payroll withholding tax or wage tax is levied on employees or their employer.[20] The tax is collected by the employer[21] through withholding from the wages or salaries due to her employees.[22] The employer needs to maintain a payroll administration as a basis to determine the correct amount of withholding tax for each individual employee. In this case, the employer had paid cost reimbursements to her employees without withholding the tax. Reimbursements of costs incurred by employees in the course and furtherance of their employer's business are in principle and under conditions not taxed, because it are business costs that should be borne by the employer. In this case, however, it was undisputed that the reimbursements were excessive and, therefore and to the extent that they were excessive, "hidden" salary subject to withholding tax. The employer is required to maintain a payroll administration, which is the basis for her payroll withholding tax return. The burden of proof that certain cost reimbursements are not taxable rests with the employer; and the employer is liable for the tax notwithstanding her regress on her individual employees. The Supreme Court accepted sampling in this case as the basis for the assessment and rejected the objection made by the Advocate-General that sampling does not allow for individualizing of employees' tax liability and, therefore, would make it impossible for the employer to exercise her right of regress. Importance in this case appears to be that the employer— regardless of whether the taxing authority has estimated the tax through sampling— has the ability to exactly calculate the tax due on the cost reimbursement for each individual employee on the basis of her own payroll administration. This is a major distinction from the passenger car tax case.

For those unfamiliar with the tax, the VAT is levied on each and every transaction down the supply chain. A business will charge the tax on each sale whether it is of goods or services (output-VAT).[23] However, businesses receive a credit for tax paid on their procurement (input-VAT).[24] At the end of the tax period, a business needs to file a return calculating its tax position by deducting the total amount of input-VAT incurred during the period from the total amount of output VAT charged during the same period. The difference is paid to or refunded by the taxing authority. Only businesses receive an input-VAT credit, so when the good or service reaches the final consumer, she actually pays the tax over the retail value. In other words, economically, the final consumer is supposed to carry the burden of the tax.[25] Legally, however, the person liable for the tax is the business making the supply of goods or services.[26]

---

20. Article 1, Wet op de Loonbelasting 1964 (Wage Tax Act 1964).
21. Article 27(2), Wet op de loonbelasting 1964 (Wage Tax Act 1964).
22. Article 27(1), Wet op de loonbelasting 1964 (Wage Tax Act 1964).
23. See for the EU, Article 2(1)(a) and (c) of Directive 2006/112.
24. See for the EU, Article 168 of Directive 2006/112.
25. See for the EU, see Preamble and Article 2 of the First VAT Directive (67/227/EEC) of April 11, 1967 OJ 14 April 1967 (L 71) 1301. It was repealed when Directive 2006/112/EC of 28 November 2006, O.J. 11 December 2006 (L 347) went into force on January 1, 2007, but the first three paragraphs of Article 2 of the First Directive were reproduced in Article 1(2) of Directive 2006/112.
26. See for the EU, Article 193 of Directive 2006/112.

Businesses are required to keep books and records from which the tax liability in any given tax period can be derived.[27] In principle, under Netherlands law, the burden of proof that the correct amount of tax has been remitted rests with the business filing the return. Statistical sampling is an adequate method to approximate VAT under- or overpayment in case of a systematic error. In the case at hand, an accountants firm erroneously did not charge VAT on its services due to a misqualification under the tax jurisdiction rules. In other words, it mistakenly believed that the services were not taxable as a domestic supply. This error itself was not in dispute. Again the argument was made that sampling could not be accepted because it does not allow individualizing of undercharged VAT to specific clients and thus would preclude the taxpayer from issuing corrective invoices to its clients. It was presented as a violation of the neutrality of the VAT system where the tax is supposed to move along the distribution chain but be a burden only to the final consumer. Where no corrective invoice can be issued, the tax would become a burden to the assessed taxpayer contrary to the stated purpose of the tax. The Supreme Court does not accept the reasoning that an assessment would only be valid if it is based on an auditing method that guarantees that the taxpayer can recharge the tax to its customers. Similar to the wage tax case, the taxpayer is certainly able to issue corrective invoices because she keeps an administration recording all sales and VAT charged. It basically comes down to the question of who has to do the actual administrative work, the taxpayer making the error or the tax administration performing the audit. The Supreme Court puts that burden on the taxpayer and rightfully so, in our opinion, since the latter has the burden of proof.

Evaluating these cases, the Netherlands Supreme Court does not accept objections against sampling based on the system of the tax, except for the passenger car tax. This raises the question of what the fundamental difference between these taxes is. There appear to be two differentiators between the car tax on the one hand and the VAT and wage or payroll withholding tax on the other hand. First, there is a difference in the allocation of the legal liability. Under the passenger car tax the person in whose name the vehicle is registered is in principle liable for the tax and the importer in his capacity as registrant of the vehicle is only paying the tax *on behalf of* the former. However the tax liability for the VAT rests with the business making the supplies of goods and services not with the final consumer, regardless the economic purpose of the tax. The liability for the payroll withholding tax is shared between employer and employee. Second, there is a difference in the ability for the assessed taxpayer to realize regress; in other words to shift the tax forward to the intended party. For VAT and payroll withholding tax the taxpayer is able to do that based on the information available in her own administration. Under the passenger car tax, the importer is unable to do this because her administration does not hold information regarding the accessories ordered by the car owner from the car dealer.

---

27. See for the EU, Article 242 of Directive 2006/112.

[C] Confidence Interval

The heart of statistical sampling is the control of risk. In estimation sampling, used in tax compliance auditing, the central question is what is the risk that the amount assessed is not accurate? Alternatively, is there risk that an assessment might unduly injure a taxpayer, or that the government cannot objectively justify the assessment? These questions will likely be fundamental to any final decision on the use of sampling where the taxpayer does not agree on methodology. Since risk can only be measured objectively by statistical sampling, there is a consensus among most agencies that statistical sampling is the only method that can successfully be used when taxpayer acceptance is not there.

But just because a statistical sampling procedure has been used does it follow that the results of the sample will stand up to legal challenge? In fact, most audit populations are subject to extreme variation (as the discussion of the zero error issue in chapter 2, clearly shows), and therefore, resulting confidence intervals will often be wide relative to the point estimate. In many cases, the only real way to narrow these intervals (provide a more accurate result) is to take samples in at least the size of several thousands of sample units, which most agencies and taxpayers will balk at. Many agencies using statistical sampling do a good job in training their auditors, and using sophisticated methods in applying, taking, and evaluating samples. But some may fail to adequately grapple with the issue of the wide confidence interval and applying the results of the sample to the assessment. Regardless of the width of the interval, some may use the point estimate as the basis of the assessment. For example, the sample shows that the true unknown tax total is likely USD 50,000, with a 90% confidence that the true unknown liability is anywhere from USD 10,000 to USD 90,000 (this is an expression of a *2-sided* confidence level). In assessing USD 50,000, a fair chance exists that the true amount will be lower and may easily be as low as USD 10,000. A taxpayer might have a fair chance that the possibility of material harm could exist. Unfortunately, in some cases when confronted with such sample results, some auditors have not disclosed the confidence interval in order to prevent any questions.

Of some note to this discussion is that a side-effect of the technology revolution is that anyone, including the taxpayer, can compute the confidence interval themselves. It is normal for government agencies, in their audit work papers, to contain the information required to make an independent evaluation. It is advisable that any audited taxpayer request these work papers and specifically the results of any statistical evaluation.

The IRS, in dealing with this sticky problem of wide confidence intervals, takes a very logical position of justification. They can, using the example of above, use the LCL of USD 10,000 as the basis of the assessment. With the same exact criteria stated in the example, it can be said with 95% confidence that the true unknown amount is at least USD 10,000 (now the confidence level is *1-sided*). Any detailed examination of such a population will very likely show that the true unknown amount is more than the amount assessed. If the burden of proof rest with the taxing authority, we believe this approach is legally the most correct one, simply because all that sampling has really proved to be underpaid in our example is the minimum amount of USD 10,000.

The question of confidence level is also subject to debate. At what confidence level are the individual rights of the taxpayer balanced with the government's obligation to come up with a true and accurate assessment? In fact, some argue that the policy of using an interval assessment based on a 95%, 1-sided confidence level is far too conservative and that individual taxpayer rights are given too much deference. It is further argued that such a policy is not really one of legal justification, but actually a policy that discourages any legal challenge in and of itself. What rational person would object to an assessment that is likely understated with such a high confidence? In response to that, there have been arguments put forward that interval assessments should use a lesser degree of confidence. The smaller the degree of confidence, the narrower the confidence interval will be. The sampling risk it is argued, while still incurred mostly by the government, is shifted a bit more towards the taxpayer, resulting possibly in a more sensible public policy. Those who believe a 95% confidence policy is too conservative are known to advocate for levels as low as 75%.[28]

The issue of confidence level may indeed ultimately be settled in a legal setting rather than by legislative fiat. The confidence levels in use today range anywhere from 75% to 97.5%. But this is not the primary legal issue surrounding the application of sampling in tax compliance auditing. At present, government agencies employ a wide range of sampling policies. Assessments can be made using non-statistical methods when more scientific approaches can easily be applied. Audits will employ probability sampling methods, but fail to statistically evaluate them and still be in compliance with agency guidelines. Amongst agencies that use statistical sampling, some use the point estimate and others the interval for the basis of the assessment. One cannot help but wonder, given this broad range of policy, what is the best and most balanced application of sampling policy in tax compliance auditing (if indeed there is one policy that works for all situations)? The courts have been historically disinterested in this issue. This question is basically left to most agencies to deal with on an individual basis.

[D] Taxpayer's Acceptance and Sampling Agreements

Legal decisions, although rare, can be found to either support or deny the government in the application of assessments based on sampling. In a few places, laws have been passed to allow the government to apply such procedures. An extreme position, which is actually common but not universal to tax agencies, that unless the taxpayer agrees to the sampling methodology, the government auditor is precluded from using *any* type of sample to estimate an assessment.

In this respect, several states in the US conclude contracts with taxpayers in regard of the sampling audit. The concept of a sampling agreement that includes the population definition for the audit, the sampling plan, an evaluation of the sampling plan and a discussion of how special situations will be handled (e.g., missing

---

28. Bright, Joseph C., Jr., Joseph B. Kadane, and Daniel S. Nagin, "Statistical Sampling in Tax Audits," *Law & Social Inquiry* (Spring 1988).

documentation, progress payments, bad debt) certainly deserves our support, because it allows to mutually address key elements and remove concerns and obstacles prior to the actual start of the audit. A good example of such an agreement is the one applied by the California Board of Equalization.

Sampling agreements fall into two categories: binding and non-binding. Binding agreements generally mean that once the sampling plan is agreed to by the state and the taxpayer, it cannot be altered in later stages of the audit. A non-binding agreement allows for modification of the sampling plan if warranted by unexpected developments in the audit.

US States that use binding sampling agreements include: Arizona, Arkansas, Florida, Iowa, Maryland, Missouri, Nebraska, New York, Ohio and Washington.

At least one state has been known to engage in a practice that at least some people have found less than ethical. This state threatened to not apply tax overpayments as an offset to tax underpayments in the sample, if the taxpayer refuses to sign the state's sampling agreement. In addition, the sampling agreement has been used in court cases as a basis for dismissing the taxpayer's claim. It is important to note that a sampling agreement does not give carte blanche to the tax authority in the use of sampling methodology. The courts should, therefore, at all times uphold a taxpayer's right to protest the use of a defective sampling plan or defective sample, because it is unreasonable and against common sense to assume that her consent would include the use of defective methodology and inaccurate execution. At the contrary, the taxpayer agreed to sampling under the assumption that it would be conducted in a manner consistent with adequate standards of professionalism.

[E]   Fairness

There is an issue sometimes levied against sampling that involves the expertise level of the taxpayer. The statistical sampling techniques are quite complex, and the ordinary person, including most taxpayers, very well may not be able to understand the procedure. Most tax consultants, tax lawyers and accountants have none or only superficial knowledge of sampling techniques. Expertise is for hire, of course, but generally at high hourly costs, which may be acceptable if sampling procedures were applied only on the largest and most sophisticated taxpayers. It is argued that it is unfair for the state to come in and do a procedure that the taxpayer might not be able to understand because it disrupts the precarious balance of fair play. Sometimes it is argued that the state is taking advantage of the unsophisticated. In rebuttal, some cases, irrespective of the level of knowledge of the taxpayer, call for sophisticated sampling procedures. But it seems illogical that the state should be precluded from performing the most appropriate audit procedure merely because the taxpayer is unsophisticated. Similarly, some may argue, it does not follow that the attorney should refrain from making an argument in court that benefits the client merely because the client does not fully understand it. Nor does it follow that the physician should not perform a necessary surgery because the patient is incapable of fully understanding that. Their immanent truth notwithstanding, these examples have no bearing in the case of a tax audit, because where the doctor and the lawyer are engaged for the benefit of their client and patient, the tax auditor represents an interest conflicting with that of the taxpayer. Nevertheless, it is unconvincing to argue that the application should be avoided of sophisticated, scientific techniques, where their use is appropriate, merely on grounds of their complexity.

But this does amplify the need for transparency and the provision of proper information to the taxpayer when resorting to sampling auditing. Moreover, the taxing authority has a genuine self-interest in explaining the sample technique it is planning to use. There is no benefit for the state if a tax payer erroneously feels "cheated out of money" regardless whether that is caused by the taxpayer's ignorance. Building rapport with and getting the buy-in from the taxpayer for the application of sampling are important factors in preventing reduced future voluntary compliance. From these perspectives, all tax authorities should give ample consideration to the concept of sampling agreements.

§3.03 EXTRAPOLATION

An interesting and vital question is whether the results of a sample test can be applied outside its direct scope or, in other words, whether extrapolation is allowed. In the Netherlands case, discussed earlier, the Supreme Court took the position that such would only be allowed if there is sufficient ground to assume that the facts in the years not included in the sampling are consistent with the sample year.[29] In this case, the principle of extrapolation was accepted, although the revenue service could only "proof" consistency in the type of error made not its frequency. In our opinion, the Netherlands Supreme Court has erred when it ruled that the results of sampling in relation to one tax year can be used as the basis for the tax assessment for previous years. A sample has only meaning for the population from which it has been taken and cannot reasonably provide guidance in terms of error for populations that are not included in the sampling audit. Extrapolation cannot be justified on grounds that the same type of mistake was made in previous years, because the frequency of the error may differ from year to year. In other words, the sample taken from the population of one specific year is not representative for previous years. In actuality, the Supreme Court accepted the application of random sampling of one year as the non-random sampling for previous years. Such an approach does not provide reliable results for these previous years because the error rate for those years is unknown. The Supreme Court conditioned its acceptance of extrapolation on consistency of the facts in the non-audited years with those of the sample year(s). However, the ruling constitutes a contraction because it is impossible to establish the required consistency without sampling the non-audited years.

---

29. Case 39.866, *Supra*, para 3.3.3. Dutch text: "... in beginsel slechts verantwoord indien er voldoende grond is om aan te nemen dat de desbetreffende tijdvakken een constant beeld vertonen wat de van belang zijnde feitelijke situatie betreft."

§3.04 INTEREST AND PENALTIES

A tax assessment is normally accompanied by interest charges to compensate the state for the loss of use of the monies from when the tax was legally due until the date of assessment. In principle, where the tax assessment itself, using sampling auditing, is found to be acceptable, the interest charges calculated thereupon must be acceptable as well. Basically, the discussion on interest payments should follow the line of acceptance reasoning applied in section § 3.02[C] in the context of confidence interval. In cases where the burden of proof rests with the taxing authority, we would strongly argue for a tax determination, and therewith a calculation of interest, based on an interval assessment based on a 95%, 1-sided confidence level. It appears to be defendable to take the midpoint position of the interval width as a basis for the calculation of both tax and interest when the burden of proof rests with the taxpayer.

Penalties strongly differ in character from interest charges. The latter constitute compensation for loss of use, the former are punishment for non-compliance. Tax penalties are generally levied as a percentage of the tax assessment and the question arises whether that is acceptable when the assessment of tax is estimated through statistical sampling. This is a very relevant issue because tax penalties vary depending on the degree of culpability. Sampling may provide a good estimate of the size of underpaid tax, but it has no apparent bearing in determining the degree of fault. Relevant in this context— at least for European jurisdictions and interesting for all other jurisdictions— is a ruling by the Netherlands Supreme Court in 1985 that the European Human Rights Convention[30] applies to tax penalties as they constitute a form of criminal prosecution.[31] As a result, a taxpayer benefits from the protection offered by the Convention such as the presumption of innocence and the right to a fair trial. Without an honest and fair assessment of the taxpayer's culpability, the innocence presumption will be violated. A taxing authority could of course simply refrain from applying a penalty or, in jurisdictions that apply a minimum penalty, limit the penalty to that minimum. But this solution, although feasible, is not satisfactory, because it removes a financial obstacle for, or perhaps even incentivizes, reduced compliance. Perhaps a separate penalty system should be devised for tax underpayment determined on the basis of sampling auditing. If the same type of error has been detected during earlier audits, then the taxpayer is obviously a repeat offender which would justify a higher penalty. In addition, sampling does provide information on the frequency of the mistake as a percentage of total entries. However, the error frequency is merely an estimate that can only be located roughly as somewhere on the line connecting the opposite sides of the interval width. What the taxing authority can arguably prove is only the lower point of frequency and the penalty should be based on that point. For

---

30. European Convention for the Protection of Human Rights and Fundamental Freedoms, Rome, 4 November 1950, 213 UNTS 222 entered into force 3 September 1953 as amended by Protocols No's 3 (ETS No .45), 5 (ETS No. 55), 8 (ETS No. 118), and 11 (ETS No. 155) which entered into force on 21 September 1970, 20 December 1971, 1 January 1990, and 1 November 1998 respectively.
31. Hoge Raad, 19 June 1985, BNB 1986/29.

example, if sampling indicates an error rate between 20% and 40%, the penalty rate applied should be the one appropriate for the 20% error rate.

A related issue is the tax amount on which the penalty should be applied. That is not necessarily the amount of assessment. We have argued afore that the tax assessment in case the burden of proof rests with the state should be at the lowest level of the interval width, because that is all the state has effectively proved to be underpaid. Clearly, a penalty levied on the lowest possible level of tax assessment would be equally fair. We have also argued that it would be acceptable for the state in case the burden of proof rest with the taxpayer to take the midpoint of the interval width for calculating the tax assessment, because there would be an equal change that in reality the underpaid tax would be higher or lower. However, we believe, that the same reasoning cannot be applied for calculating the penalty, because in the context of a criminal prosecution the burden of proof for the correct application of a penalty will at all times rest with the state. Therefore, the penalty should still be calculated on the basis of the lowest possible level of tax underpayment. To clarify our reasoning, we return to the example used above. Assume that the sampling shows that the true unknown liability is anywhere from USD 10,000 to USD 90,000 with the midpoint at USD 50,000. Further assume that the penalty rate for these amounts would be 10%, 50% and 25% respectively and the interest rate is 5%. In case of the burden of proof resting with the state, the tax assessment would be USD 10,000 and the penalty would be 10% of that amount, or USD 1,000 with interest being USD 500. If the taxpayer has the burden of proof, the amount of underpaid tax would be assessed at USD 50,000, however, the penalty level would remain at 10% calculated over USD 10,000, thus being USD 1,000, and interest would be 5% of USD 50,000, or USD 2,500.

§3.05 CONCLUDING REMARKS

Statistical sampling does not belong to the standard toolbox of lawyers, and their unfamiliarity with the subject matter may explain why lawyers— particularly judges— tend to abstain from involvement in arguments concerning technical applications surrounding sampling. Unknown as the territory may be, some exploration by lawyers will be required if they wish to assist taxpayers in their audit defense with any chance of success or render judgment on a sampling dispute while sitting on the bench. Informed opponents— i.e., taxpayers and their advisers— are also in the interest of the taxing authority, because it will ultimately improve the quality of the sampling audits. Moreover, a taxpayer who does not understand the methodology underlying her tax assessment is more likely to revolt through future reduced voluntary tax compliance. The goal of a tax compliance audit is a true and accurate assessment. Given that very large populations are commonly confronted by auditors, *and* the above discussion of audit error in examinations of such populations, it does not follow that a complete audit of all of the items in the population will necessarily be the optimal course of action in every case. Some sort of sampling will be called for and possibly provide a truer or more accurate conclusion in at least some audits. Allowing sampling audits, therefore, seems to be both rationally consistent and practical. In this chapter we have closely examined the permissibility of sampling from different angles. The allocation of the burden of proof plays a nuanced role. In our opinion sampling should be allowed regardless of whether the burden of proof rests with the taxing authority or the taxpayer. However, the allocation of proof has an influence on the amount of tax assessment. Where the burden of proof rests with the state, the assessment should be based on the minimum error position given the width of the confidence interval. That is fair, because that is all the state has actually proven to be wrong, irrespective the fact that the amount of underpayment is likely to be higher. On the other hand, if the burden of proof rest with the taxpayer, a tax assessment calculation based on the mid-range of the confidence interval (point estimate) appears to be fair and appropriate. Interest charges should be calculated on the same basis. However, we do not apply the same rationale to penalty charges. Penalties will vary depending on the degree of culpability, which cannot satisfactory be assessment through sampling. Without a fair assessment of the taxpayer's culpability, the innocence presumption will be violated. Perhaps a separate penalty system to be applied in case of sampling, where the penalty percentage depends on the rate of error, would be a good idea. Since the penalty is calculated as a percentage of the tax determined to be due, the point of assessment is relevant to judge the fairness of the penalty. We have argued that it would be acceptable for the state in case the burden of proof rest with the taxpayer to take the midpoint of the interval width for calculating the tax assessment, because there would be an equal change that in reality the underpaid tax would be higher or lower. Nevertheless, the penalty rate should be set at the rate applicable to the lowest possible level of tax underpayment.

The type or system of tax is, in principle, an irrelevant factor in determining the permissibility of sampling audits. Even in systems that basically use businesses (under VAT) or employers (under wage tax) as unpaid tax collectors from an economic perspective, sampling is acceptable because these businesses and employers have a direct liability for the tax and are able to shift the tax based on information available in their own administration.

Extrapolation, i.e., the extending the results of sampling for one year to determine the tax liability for previous years, cannot be accepted. We strongly disagree with the ruling of the Netherlands Supreme Court on this matter. Although the Court conditioned its acceptance of extrapolation on consistency between the facts in the years not included in the sampling and the facts in the sample year, the reasoning is logically flawed the required consistency can only be established by sampling the non-audited years as well.

Rational thought would seem therefore to say that sampling can be used in auditing, irrespective of the taxpayer's acceptance of such methods. However if one accepts this premise, it does not follow that *any* sampling procedure applied is justifiable. Where it can be shown that the auditor used poor methods, it would seem that a taxpayer could rebut an assessment. That should also hold when a taxpayer is party to a sampling agreement with the taxing authority.

§3.06 REFERENCES

Bright, Joseph C., Jr., Joseph B. Kadane & Daniel S. Nagin, "Statistical Sampling in Tax Audits," *Law & Social Inquiry* (Spring 1988).

Zwemmer, Jaap W., Annotation Hoge Raad 2008, no. 40.806, 39.866 and 40.474, *NTFRB*, 2008-24.

Part III
Technology Use in Government Tax Administration

CHAPTER 4
Technology and Taxation in Developing Countries

Richard M. Bird and Eric M. Zolt*

§4.01 INTRODUCTION

Technology has influenced the way we work, play, and interact with others. It is not surprising that technology has also affected how tax systems are designed and administered in developing countries. These changes have not always been for the better. In a pioneering study of tax administration in developing countries, Radian (1980) noted that the three decades since World War II had seen a number of cycles of ineffective reform, including computerization. Many countries shared the experiences of Trinidad, in which the Commissioner of Internal Revenue said that "since 1969 we have not produced any meaningful statistical data. In that year, we transferred our returns, processing and accounting work onto a computer" (Radian 1980, 217).[1]

Technological change continues. Most countries have now moved from rooms full of clerks posting entries by hand in large ledger books to widespread use of computers to administer their tax systems.[2] The transition from hand to mouse has been incomplete and uneven. Major differences exist among and within developing

---

* This is a revised and updated version of "Technology and Taxation in Developing Countries: From Hand to Mouse," *National Tax Journal*, 61(4, Part 2) (December 2008), 791-821.
1. Radian (1980, 224) went on to conclude that "all four countries studied [Trinidad, Jamaica, Thailand, Philippines] have succeeded in installing a computer, but not one managed to use it properly.... If the data input is poor, the processed output cannot be better. This first principle of computer processing has been too frequently ignored by those who advocated the use of computers and tried to sell them to poor countries."
2. In contrast to the 1960s, this second generation of computerization relies on personal computers and a network-based information system, rather than stand-alone mainframe computers. Although many tax systems in developed countries are now essentially web-based, few developing countries have adopted web-based systems.

countries, both with respect to how their tax systems are designed and administered, and, more generally, with respect to how technological advances have changed the manner in which their economies operate.

Roller and Waverman (2001) demonstrate that the introduction of mobile telephones has enabled developing countries to bypass the heavy infrastructure development of land-based telephone systems, and has facilitated market integration and more rapid economic development. Does the use of technology in the tax systems of developing countries mark a similar opportunity for developing countries to improve tax administration and design? Ideally, to answer this question, one needs to consider the costs and benefits of different types of technological changes for administrators, taxpayers, and third parties involved in the taxing process. We cannot undertake this major task here, however. Instead, we present an overview and selective survey of the issues raised by how technology affects tax administration and design in developing countries. Technology is not a 'magic bullet' to solve the manifold problems of development taxation (Bahl and Bird 2008). However, it may provide part of the answer for many countries.

For example, one part of the answer focuses on how tax administrators can use technology. Both the size of the tax collection agency (Slemrod and Yitzhaki 1987) and how the tax agency allocates its budget among its different functions of enforcement, return and data processing, and taxpayer service (Plumley and Steurele 2004) may be affected. What are the necessary preconditions in developing countries for successful adoptions of new technology?[3]

Another part of the answer focuses on the larger question of tax system design. Slemrod (1990) notes that the design of optimal tax systems requires consideration not only of changes in the technology of collecting taxes, but also of how technology may alter the economic environment in which governments seek to collect revenue.[4]

---

3. Jenkins (1996, 13) noted in an early review of technology and taxation in developing countries that "if we want to avoid the cycle of unfulfilled expectations, we need to have a clear strategy for administrative reform, which is much broader and more sophisticated than one which simply implements information technology." Similarly, in words echoing Radian's work a quarter century earlier, Dhillon and Bouwer (2005) noted that imposing new information technology (IT) systems without altering the underlying business processes of the tax administration, or without establishing sufficient links to information providers within and outside the public sector, as well as providing adequate staffing infrastructure, was a recipe for failure.
4. Technology may influence the institutional and political context in many ways. For instance, technology may change the tax environment by altering distribution methods or reducing cash transactions. Technology may improve the quantity and quality of information available to taxing authorities and their ability to use that information effectively. Technology may make tax administrations more effective by improving information flow, facilitating coordination, and improving their allocation of resources. Technological changes may reduce taxpayer compliance costs by improving information and services to taxpayers (e.g., software for maintaining books and records, and for calculating tax liabilities, or electronic or return-free filing alternatives). Technology may reduce opportunities for corruption by reducing the face-to-face interaction between taxpayers and taxing authorities. On the other hand, since silver linings seldom arrive without clouds, technology may equally well increase corruption by increasing the opportunities for more sophisticated collaboration between taxpayers and corrupt officials.

Policymakers need to consider the impact of changes in technology on both the design of specific taxes and the relative use of different tax instruments in raising revenue.

We begin by examining how technology may be used to improve tax administration. We then turn to a broader and less explored question, the potential for technology to change the design of tax systems in developing countries. Following a brief review of some issues common to both developed and developing countries that result from the increased use of technology in tax design and administration, we conclude by offering a few tentative observations about the future relationship of tax and technology in developing countries.

§4.02 TECHNOLOGY AND TAX ADMINISTRATION

Tax advisors frequently note that "tax administration *is* tax policy" in developing countries (Casanegra de Jantscher, 1990, 178). Limitations in tax administration constrain tax policy choices. In this section, we first consider several challenges facing tax administrations in developing countries. What tax administrations actually do is not so much collect money directly, but rather collect the information needed to administer the revenue laws and then use that information as efficiently and effectively as possible. In the next section we therefore consider how technology may improve the ability of tax administrations in developing countries to perform these tasks.

[A] *Challenges Facing Tax Administrations*

A large literature has developed examining the difficulties facing developing countries in administering tax systems (Bird 2004). Some of the key challenges are reviewed here. The first two challenges— the size of the agricultural and informal sector and the use of the financial sector— relate to the economic environment in which tax administrators operate. The other three challenges we review— organizational change, administrative capacity, and political will— relate to the ability of tax administrators to improve their efficiency whether through technological or other changes.

[1] *Size of agricultural and informal sector*

Many developing countries have a large traditional agricultural sector and a significant informal (shadow) economy, both operating largely outside the formal tax system. No country has managed to tax these sectors effectively (Alm, Martínez-Vazquez, and Wallace 2004).[5] As a result, the tax base that tax authorities can potentially reach is relatively small in many developing countries.

---

5. The relative size of the informal sector is often three or four times larger in developing countries as compared to developed countries (Alm, Schneider, and Martinez-Vazquez 2004). The size of the untaxed economy may itself be a function of the design and implementation of the tax system. For example, the high social insurance tax rates levied by some countries create an incentive for a large informal economy by discouraging employers from reporting the extent of

The conventional wisdom has long been that the informal economy consisted largely of small providers of goods and services operating largely outside the formal economy. There was also a belief that as countries developed economically, the size of both the agricultural sector and the informal economy, relative to total economic activity, would decline. However, as Chen (2005) highlights, in some countries not only is the informal economy operating in almost all parts of the economy but there is also substantial interaction between the formal and informal economy. Not only do many firms in the formal sector buy from and sell to informal firms, they may even control (or be controlled by) 'shadow' enterprises. Nor does economic development necessarily result in a smaller role for the informal economy. Indeed, when the barriers to formalization are as high as they are in many developing countries, growth may be more likely to manifest itself in the informal sector of the economy.[6]

The decision to operate in the informal economy rests on the relative costs (e.g., tax and regulatory costs) and benefits (e.g., access to financial sector and legal systems) of operating in the formal economy. These relative costs and benefits vary between developed and developing countries, as well as among developing countries. One important cost of operating informally is the probability of detection by tax or other government authorities, and resulting penalties. Here, again, countries differ greatly. By providing additional tools to observe and monitor transactions and taxpayers, technology may significantly enhance the ability of tax authorities to detect economic activity in the informal sector. For example, technological improvements may provide tax authorities with greater capabilities to track use of physical inputs, electricity, and labor and, hence, to estimate revenue and profits.

[2]   Use of the financial sector

Countries also vary greatly in the role played by financial institutions in the operation of the economy.[7] Where activity is primarily in cash or barter transactions, it is difficult to monitor transactions.[8] In contrast, the use of banking channels for payment makes transactions easier to observe and monitor. The growth of the financial sector and its greater role in the market economy broadens the potential scope of taxation and makes administration of certain taxes easier.

The increased use by businesses of financial institutions to channel receipts and expenditures provides information not only on those businesses, but also on their

---

employment and encouraging the under-reporting of wages (Rutkowski 2007). The resulting lower tax revenues often lead governments to raise tax rates still further, thus exacerbating incentives to evade taxes.

6. World Bank (2008) attempts to assess the height of these barriers, which were originally highlighted by the seminal work of DeSoto (1989). Stern and Barbour (2005) examine the extent to which tax barriers have deterred small business growth in sub-Saharan African countries. Bennett and Estrin (2007) emphasize the importance of informality as a stepping stone to growth.
7. The importance for taxation of the development of the financial sector has been emphasized recently by Gordon and Li (2009).
8. In recognition of this problem, India, for example, allows the deduction (for income tax purposes) of only 80% of current expenditures (above a small amount) that are paid in cash.

suppliers and on businesses further down the economic activity chain. Similarly, the increased use of credit cards, or more sophisticated electronic payment mechanisms, provides not only information as to the financial capacity of the purchaser, but also information that may be used to confirm the value-added tax (VAT) and income tax filings of the sellers.[9]

[3]   Organizational change

It is difficult to conceive of how a modern revenue administration could perform its tasks effectively and efficiently without making extensive use of information technology (IT). In many developing countries, however, the expectation that new IT alone will result in vast improvements in revenue administration has not materialized. Three reasons for such failures may be singled out:

(1) the successful introduction of new IT generally requires the complete rethinking and redesigning of administrative systems and processes;
(2) success in reorganizing to use new technology requires close attention to the human dimension of revenue administration. Even the best IT system will not produce useful results unless there are real incentives for tax administrators to use the system properly; and
(3) the successful introduction of IT requires countries to make good decisions as to the manner and extent key IT components can be outsourced to private contractors.

Nonetheless, it is clear that radical improvement in tax administration requires changes in organization and methods, and modern IT greatly facilitates the needed transformation (Engelschalk, Melham and Weist 2000).[10] To take an example from some years ago, a 1992 study of the enforcement efficiency of the income tax department in India identified the following problems: poor use of information collected by the central intelligence branch; ineffectiveness of surveys of business premises; absence of an adequate system of taxpayer identification numbers; absence of an adequate system of third-party information collection; and deficiencies in the record-keeping system (Das-Gupta, Mookherjee, and Panta, 1992). Although many of these problems can be

---

9. For example, since 1999, Korea has encouraged the use of credit cards through two incentives. First, 20% of credit card expenditures can be deducted from the user's taxable income. Second, a lottery was introduced in 2000 to further encourage credit card use. Although credit card use has increased rapidly in recent years, it is not clear to what extent these incentives may be responsible.
10. For experiences of particular countries with administrative reform, see Bird and Casanegra (1992), Jenkins (1996), and Das-Gupta and Mookherjee (1998). A particularly interesting account of business process re-engineering in Singapore is Sia and Neo (1997). Pinhanez (2008) examines in detail how the nature of the VAT (the principal tax of Brazilian states), shaped IT, organizational redesign, and human resource policies in their tax administrations.

solved by adopting appropriate and available technology, successful use of IT requires restructuring and retraining the tax administration.[11]

As with most new technologies, IT is a double-edged sword. In the hands of taxpayers, it may make tax administration more difficult (especially in an open economy). In the hands of tax authorities, it may enable a more robust response to such challenges. Experience suggests a number of lessons with respect to the successful application of IT in tax administration. An appropriate strategy must consider the obstacles and constraints arising from such organizational rigidities as civil service salary structure and procedural hurdles in acquiring the necessary expertize, hardware, and software.[12] Equipment and software should be standardized to facilitate training, operation, networking, and maintenance. Whenever possible, software should be bought off the shelf rather than developed internally, both for cost reasons and to accommodate subsequent technological developments. Ultimately, the pace of change and the success of any modernization program depend on human resources—on the training and skills of the people who are expected to use and operate the technology.

Considerable organizational re-engineering is usually needed to gear the tax administration to a computerized environment.[13] In certain cases (e.g., with property taxes in Indonesia) it may be advantageous to reorganize tax administration by sector (Kelly 1996). In other cases it may be better to follow the Spanish system, where key information regarding a taxpayer's obligations (e.g., the filing of returns and the making of payments) is combined with a tax 'vector' created for each taxpayer (Moya and Santiago 1992). As Pinhanez (2008) explains, the very nature of the VAT requires administrators to understand business 'chains,' the interconnected nature of activities in different stages of production and distribution process.

Experience in Kenya and elsewhere demonstrates that the successful introduction of new technologies requires consideration of the susceptibilities of existing staff

---

11. A striking illustration of the difference in the pace of new technology, and the change in human resources required to meet effective use of that technology, occurred when Indian tax authorities seized a large volume of computer-based records in 1992 in connection with an investigation. The clerical staff then proceeded to punch holes in the seized floppy diskettes in order to attach identifying tags. In recent years, India has moved much further along the road of adopting appropriate technology in tax administration (UN 2007), although, from different perspectives, both Kumar, Nagar and Samanta (2007) and Manglik (2008) suggest that there is still a long way to go.
12. For an overview of how IT fits within an administration reform strategy, see Gill (2000). Pinhanez (2008) illustrates and elaborates this point with respect to Brazilian state tax administrations.
13. Singapore has been one of the most successful countries in using technology to modernize and improve its tax system. (Jenkins 1996, Das-Gupta and Mookherjee 1998, and Bird and Oldman 2000). This reform demonstrates the importance of careful and systematic planning and the thorough re-engineering of all the business processes of the tax administration. Singapore's experience also suggests that even good tax administrations may often gain from outside assistance in undertaking extensive IT reforms. Even then it is a long process: Singapore began its conversion to an IT-based system in 1992 and the process was still going on eight years later in 2000 (Bird and Oldman 2000). The Brazilian state tax reforms discussed in Pinhanez (2008), however, suggest that motivated and well-led revenue administrations, even in relatively poor states, can substantially improve their performance through appropriate use of technology within a relatively short time.

and their resistance to change (Peterson 1996). Indeed all those in a position to affect how well any new IT system can function must work together.[14] As a complex system is more likely to engender resistance and problems, the design, structure, and operations of the system, should be as simple as possible. In some situations (such as in Mexico) it may even be advantageous to entrust part of the responsibility for setting up an information system to organizations outside the tax administration, or even outside the government.[15] We return to the question of privatization later in this chapter.

[4]   Administrative capacity

Many who examine tax administration in developing countries conclude that there is simply insufficient administrative capacity, usually defined in terms of skilled human capital, for the tax administration to function properly. The problems encountered in countries that have undertaken large-scale IT modernization projects include the following:

- cost and time overruns, often resulting from overambitious plans and poor project management;
- inadequate flexibility in learning from experience and changing appropriately;
- lack of independent oversight (one cannot expect independence when the careers of those responsible are being determined by what is being supervised);
- either the 'NIH' (not invented here) problem— over-emphasis on the uniqueness of the situation and unwillingness to learn from the experience of others— or its opposite, the 'copycat' problem— if it worked there, it should work here;
- lack of attention to complexities and delays that arise from such basic issues as procurement regulations and staffing policy; and
- inadequate attention to the effects on the ultimate 'client'— the citizen-taxpayer.

Such problems are not unique to developing countries.[16] Indeed, even in countries with very low levels of administrative capacity, technological innovation may to some

---

14. In one country, for example, an expensive new computer system was completely disabled the day after it was launched simply because an employee looking for a convenient place to plug in his tea kettle disconnected a key power cord at a critical moment.
15. There are limits. In one country, a foreign service provider proposed, for security reasons, to install the servers at its home country's embassy. For an early but still useful consideration of outsourcing in tax administration, see Ramirez Acuña (1992).
16. For example, a recent audit in Canada found one large IT project to have been a complete waste of money. The audit found that largely because of inadequate communication between the developers and the intended user, the final product was completely rejected by the relevant operational division of the CRA, which then undertook a new project on its own to better meet its needs (Auditor-General 2008).

extent be able to substitute for inadequate human capacity. For example, while a few highly skilled people would still be required to implement and operate any modern data processing system, it might be easier in many countries to find, say, three capable university graduates rather than the many literate and numerate high school graduates who might otherwise be needed to do the same work.

More generally, at any given level of administrative capacity, more can be done better with appropriate advanced technology, often by complementing and increasing the productivity of skilled staff instead of replacing it. In some cases, technology can and has substantially extended the capacity of tax administration officials by permitting them to assemble and evaluate the mass of information already currently available but not effectively used. Technology alone cannot do the job of good tax administration, and good tax administration can be carried out without technology. Technology increases the opportunities with respect to what can be done in any tax administration and often makes it possible to perform administrative functions both differently and better than without technology.

[5] Political will

Although technology can improve tax administration in developing countries, some countries lack political will to use technology effectively to improve tax compliance. When the will is there, the way usually already exists and need only be introduced. Undoubtedly, introducing creative technology in many developing countries will yield significant gains, as shown by the use of financial networks operated through mobile telephone networks to bypass infrastructure efficiencies. For the most part, what really needs to be done to improve tax administration in developing countries is well-known and can sometimes be implemented within a surprisingly short time span, as has been demonstrated in both Brazil (Pinhanez 2008) and Chile (Toro 2005). Technology may enable countries to leap over infrastructure gaps and even to overcome (to some extent) human capital deficiencies. But it cannot circumvent the critical political obstacles that plague tax administration in many developing countries.

[B]   Using Technology to Improve Tax Administration

Tax administrators in developed and developing countries play many roles. They are expected to collect revenue, process returns and information, limit tax evasion, provide services to taxpayers, and in many countries, implement social programs through the tax systems. The objectives and policies of tax administrators differ among countries and over time. Yet, without a clear understanding of the short-term and long-term objectives, it is difficult to measure the success or failures of tax administrators. It is also difficult to decide how to allocate resources, whether personnel or investments in technology, among the different administrative functions. The IRS has found it hard enough to make the sound, informed decisions on such matters (Plumley and Steuerle 2004); it is not surprising that the task has proven even more challenging in the difficult circumstances facing developing countries.

Like any other government agency or private sector enterprise, tax administrators need to make difficult choices on allocating scarce resources among different types of taxes, different administrative functions, and different types of technology. To make informed decisions on alternative feasible technology investments, for instance, tax administrators need estimates of the current costs of administering the tax system as well as the costs of administering particular taxes and the expected costs and benefits from the additional investment.[17] Ideally, decision-makers need not only total or average costs but also the relative marginal costs for different investments and different tax instruments. Even with this information, decision-makers will have different time frames and objective functions. For example, if the focus is short-term revenue, they may choose not to invest in projects with potentially large long-term benefits and the potential to strengthen the state by improving its 'tax interface' with society (Moore 2007) may be ignored. Such subtleties may be left aside in most developing countries, which lack the relevant information to make such determinations. In many such countries, the only data systematically collected by tax administrations are current collections. Even the large revenue administration improvement programs funded by international agencies such as the World Bank have not, to our knowledge, assembled or used data on relative marginal costs and benefits to make informed decisions on different proposed administrative reforms. In this, as in other areas of development policy, faith and ideas derived from experience elsewhere appear to have determined more allocative decisions than evidence-based analysis.

The lack of hard evidence does not mean that there has been no change in the organization or functioning of tax administrations in developing countries. On the contrary, there has, for example, been a substantial movement away from organizing tax administration on a tax-by-tax basis towards organization based on administrative functions (Vehorn and Brondolo 1999) as well as special attention in many countries to dealing with the large taxpayers who provide most of the tax revenue (Baer, Benon and Toro 2002). In addition, some countries have introduced more 'autonomous' revenue authorities.[18] Here, however, we focus primarily on the question of how technology has the potential to change how taxing authorities perform the interconnected tasks of locating and identifying taxpayers, information reporting and withholding, processing returns, auditing, collecting, educating taxpayers, and providing taxpayer services.

---

17. Costs include not only the costs of the tax authority, but also the costs imposed on taxpayers and other parties. (Slemrod 1990). As the review by Evans (2003) makes clear, almost no such compliance cost data is available in developing countries. For an interesting early review of the allocative problem that stresses the importance of objectives other than simply revenue, see Shoup (1969, chap. 17).
18. For detailed discussions of such semi-autonomous revenue authorities, see Gray and Chapman (2001), Fjeldstad (2002), Mann (2004), and Taliercio (2004). Our reading of this growing literature is that any country able to assemble the will, strategy, and resources needed to reform substantially its tax administration, probably does not need such a separate revenue authority; countries lacking these critical ingredients are unlikely to be successful even if they create such authorities.

The availability, cost, and accessibility of computers make them ideal for the large-scale information-processing and coordination problems facing tax administrations in even the poorest countries.[19] Among the areas that may be computerized are: (1) taxpayer records and tax collection (taxpayer compliance); (2) internal management and control over resources; (3) legal structure and procedures; and (4) systems to lower taxpayer compliance costs.

[1] Tracking taxpayers

Almost all tax systems use a taxpayer identification number (TIN) to track taxpayers.[20] In every country that has successfully adopted improved technology for tax administration, allotting a unique identification number has been a necessary requirement. Without such a number, information can neither be stored properly nor used effectively. Countries may use a number unique to the tax system or one linked to other government activities. Several countries have begun issuing smart ID cards to citizens that contain TINs as well as other information.[21]

Improvements in technology allow governments to coordinate the numbers assigned with respect to various government services and financial services to TINs issued by taxing authorities. The coordination will make it more difficult for those without TINs to access government services to obtain passports or driver's licenses, register cars, transfer and register property, or use public schools or hospitals. TINs could also be required to open bank accounts, purchase airline tickets over certain dollar amount, or gain access to electrical, gas or water services— thus increasing the costs of operating outside the tax system.[22]

---

19. Tax administrations need more expertise in information technology in part because many of their largest taxpayers (multinational companies and large domestic firms) employ sophisticated computer systems. We do not discuss here the important questions raised for taxation by cross-border digital commerce: for a preliminary look, see e.g. Bird (2005).
20. Good tax administration requires a reliable single, centrally maintained register of taxpayers. This register should contain only such relevant particulars as the name, address, and the nature of business or activity of the taxpayer, together with location of branches, each under a unique taxpayer identification number (TIN), which should remain permanent and should not change, for example, when taxpayers change location.
21. Such cards may, for example, contain a microchip which has a digital image of the user's fingerprint. To use the card for commercial or tax transactions, the user presses his or her finger against a sensor on the card. The sensor reads the fingerprint and compares it with the fingerprint stored on the card's memory chip to ensure a match (Ainsworth 2006). As an illustration of the variability in the use of technology within developing countries, one of the authors was recently in a country that required digital fingerprinting and verification to get a boarding pass and to board an airplane. That country does not have reliable TINs.
22. Werbach (2007) and Eschet (2005) provide (as yet very limited) examples of applications along some of these lines in the US and UK. To some extent, of course, such applications may be considered simply an extension of the tax compliance certificate some countries currently require before e.g. issuing a passport. Such programs may have increased tax compliance to a limited extent; but in some instances they have also demonstrably provided an occasion for extortion and harassment as well as raising yet another barrier to moving from the formal to the informal sector. China is reportedly testing a program that combines smart identification cards with video surveillance, thus allowing officials to track movements of card holders (Bradsher 2007). George Orwell would presumably be horrified, although not surprised. For a review of

[2] Information reporting and withholding

An important task of tax administration is to bring together information from different sources, both within the administration, and from other relevant government and private sources, in order to verify the information supplied by taxpayers themselves.[23] Tax laws in most countries already require various private and public agencies to furnish information regarding various transactions and activities to the tax authorities. In some, but not all cases, those agencies are also supposed to withhold a part of the payment made by the agent to the potential taxpayer. Withholding thus serves the two-fold purpose of helping to identify potential taxpayers and ensuring that at least a part of the tax is realized at source, thereby minimizing risk as well as delay in payment. Neither internal nor external sources of information are of any use in the absence of an efficient system of monitoring, or of adequate IT infrastructure to collate and store data with easy access for retrieval and cross-checking. A reliable single, centrally maintained register of taxpayers, each with a unique TIN, is therefore essential.[24]

Withholding in developing countries could cover not only traditional items such as wages, interest, and dividends, but also professional fees, payments to independent contractors, rents, and (in some instances) a wide range of business transactions. Some countries have even introduced what may be called reverse withholding in which purchasers (government agencies or large enterprises) withhold tax from sellers (small enterprises). Such widespread withholding is not a panacea (Soos 1990). It makes its own information demands as the tax administration must be able to control withholders to make sure they hand over to the Treasury the amounts withheld, and it must also be able to check whether amounts that taxpayers credit against their liabilities have in fact been withheld. But it can still be very useful, particularly with respect to imposing some taxes on the informal activities.

From an administrative perspective, most taxes collected in developing countries come from a relatively small number of tax collecting agents.[25] Accurate tracking of

---

efforts by British authorities to monitor Orwell's movements, see http://books.guardian.co.uk/news/articles/0,,2161853,00.html. We discuss the issue of privacy below.

23. Useful sources of third-party information may include: financial institutions for transactions such as bank loans, bank drafts, or credit cards; insurance agencies for assets insured; public and semi-public agencies granting various licenses and permits for imports, exports, the existence of businesses or trade; or the construction of houses; utility authorities for connections of electricity, water, or gas; agencies operating communication services, telephone, or cable television service; professional societies for persons enrolled as doctors, lawyers, or accountants; authorities or agencies empowered to register transactions in real property; and even newspapers carrying advertisements for sale of goods and services, and real estate. In some developing countries, tax administrators have told us that it is often easier to get such information from private organizations than from other public sector agencies. In the public sector, control over information is power and agencies jealously protect their power.

24. Even countries with a very high-level IT capacity, like India, may fall short in this respect. For a detailed account of the problems with the TIN used for income tax purposes in India, the Permanent Account Number (PAN), see Manglik (2008).

25. Examples of tax collecting agents are customs administration (VAT and excises on imports, import surcharges, and tariffs), social security agencies (social security contributions and personal income tax (PIT) on transfers), the government itself (PIT withholding on wages),

fiscal flows through such large entities, which probably account for 80% or more of current collections in many countries, is critical to successful tax administration.[26] Before devoting much effort to this difficult task, however, it is critical to ensure that tight control is maintained over the payments and liabilities of large taxpayers. One way to do so, commonly recommended by experts, is to set up a large taxpayer unit (LTU) to monitor closely the non-filing, stop-filing, and compliance behavior of such taxpayers (Baer, Benon, and Toro 2002). In some developing countries, experience in computerizing the information flowing through and to such LTUs has proven to be a useful testing ground for developing systems that can be later extended to the whole taxpaying population.

Given the importance of the VAT as the primary source of revenue in many developing countries, the VAT is likely the tax that has the greatest potential for substantial gains from the use of technology. A key tool in enforcing VAT compliance is to cross-check purchases of one taxpayer against sales recorded by others.[27] Few developing countries have systematically used such programs to detect under-reporting, let alone outright fraud. In one country, tax authorities finally began systematic limited cross-checking only when a review of companies from whom the tax authorities had acquired services (on which it paid VAT) revealed that very few of these companies had bothered to report such sales on their VAT returns.[28] On the whole, for the immediate future less technologically advanced countries are likely to achieve greater gains improving training of auditors and adopting audit practices such as exchanging information between income tax audits (e.g., transfer pricing audits) and indirect tax audits than by depending on computer-assisted techniques.[29]

---

state enterprises (PIT withholding, VAT, excises, and corporate or enterprise income taxes (CIT)), and a few large private enterprises (PIT withholding, VAT, excises, and CIT plus perhaps taxes on dividends and interest).

26. No elaborate universal TIN system is needed for this purpose: the aim of TINs is more to extend the reach of the tax system from the existing central core of large taxpayers into the remainder of the potential tax base. It is perhaps worth noting that an effective VAT system may require an additional VAT registration number (or perhaps simply an extended TIN) because possessing a VAT number in effect gives a taxpayer the ability to issue invoices that can be used to claim VAT credits and even refunds. Developing countries have enough trouble controlling their tax system without turning over the keys to the Treasury to everyone with a TIN. This is one important reason why a relatively high VAT threshold is often a good idea in such countries (Bird and Gendron 2007).

27. Cross-matching invoices under a VAT system is similar to conditioning deductibility under an income tax system on the availability of reliable information from the recipient of the payments for which they claim deductions. Thus, similar information systems could be useful in administering income as well as consumption taxes. Ebrill et al. (2001, 148-50) present the usual case against relying on such matching. For a more positive appraisal, see Jenkins, Kuo, and Sun (2003, 179); see also Das-Gupta and Gang (2003). China is currently introducing a centralized e-invoice system controlled by the tax authorities in an effort to reduce VAT fraud.

28. Early attempts in Korea at widespread matching of VAT invoices failed to produce successful results (Choi 1990). In countries such as Taiwan and Singapore, subsequent experience in using modern information technology to match invoices of buyers and sellers seems to have been more successful. However, these countries are substantially more technologically advanced than most developing countries.

29. However, such countries can still make good use of simplified information technology to improve outcomes. For example, returns might initially be scanned to determine whether their mark-up ratios (or other parameters such as wage bills) fall within normal ranges for

The European Union (EU) has implemented a test program that allows businesses and individuals to participate voluntarily in a new fully digital VAT system, which could eventually be linked to smart ID cards.[30] The EU's Digital Sales Directive provides for a paperless VAT reporting and payment environment for non-established businesses selling to final consumers in the EU. In a somewhat similar manner, the Streamlined Sales Tax in the US under the certified service provider (CSP) model allows businesses to enter a paperless world of retail sales tax compliance (Ainsworth 2006).

[3] Processing returns and payments

One of the first uses of IT was to process tax returns and payments. Partly because banks had more adequate data processing systems than tax administrations, several Latin American countries initially outsourced the receipt and processing of tax returns and payments to the banking systems. More recently, even countries like Panama and Paraguay[31] have adopted electronic filing, which has facilitated return processing. Since 2001, Chile, likely the most advanced tax administration in Latin America, has supplied most wage-earners and pensioners with pre-populated returns that contain taxpayer identifying information, details on gross income received from various sources, tax withheld, information on certain deduction items, a calculation of the tax assessed, any credits, and the tax payable or refundable (OECD 2006).[32]

In many developing countries, however, most wage earners do not file returns at all, because their income tax obligations are considered to be fully satisfied by the tax withheld by their employers. Even in such developed countries as the UK, most personal income taxpayers do not have to file returns because cumulative averaging results in the correct amount being withheld. In 2004, when Denmark first deployed its nicely named 'no touch' (TASTSELV) tax system, 4 million taxpayers (86% of total) in 2004 received a 'pre-filled' tax return based on their digital files. Many of these returns (2.5 million) required no further action from the taxpayer because the account was fully settled. Of the remaining 1.5 million, about 1 million also accepted the official 'annual settlement' based on the data provided by the tax department. Only 430,000 taxpayers, or less than 10% of the taxable population, made any kind of corrections to

---

    comparable firms. Those that fall outside these parameters should then be subjected to additional desk investigation (e.g., cross-checking information with customs and income tax for the period in question).

30. As Ainsworth (2006) notes, there is more scope for such methods with respect to business-to-business (B2B) transactions than with respect to final sales to consumers (B2C). OECD (2005) discusses a variety of ways in which new technologies are being used in tax administration in developed countries; see also Goolsbee (2004).
31. In Panama, 36% of corporate and 24% of individual income tax returns were filed electronically in 2006. Even in Paraguay, which is considerably less advanced, an increasing number of returns are being filed through the use of official software and the Internet, although the numbers are still small— about 20,000 in 2006.
32. In 2005, 1.2 million of Chile's 1.7 million taxpayers received a complete pre-populated return over the Internet, and 57% of these taxpayers accepted this return without adjustment. In Chile, 96% of taxpayers filed returns over the Internet.

the pre-filled tax return.[33] Other countries, such as Australia, are also considering return free systems that relieve most taxpayers from the obligation of filing a return.[34] Such systems have obvious attractions for both taxpayers and the tax administration. However, questions may be raised about the long-term desirability of freeing citizens from the painful obligation of explicitly facing up to the fiscal needs of the state at least once a year.

Whether or not returns are filed with tax offices, there is much to be said for requiring payments to be made to financial institutions if only because keeping cash out of the tax office is one way of reducing opportunities for corruption. Of course, payments may be made electronically no matter how returns are processed. Canada, for example, introduced Internet filing for corporations on a voluntary basis in 2002 and since 2010 has made it mandatory for those with gross revenues exceeding CAD 10 million. Corporations (like individuals) may also pay online through arrangements with the major banks. Even in very low income countries, given the rapidly spreading use of mobile telephones as a way of making financial transactions, more use of electronic filing and payment of taxes appears likely in the near future.

[4]   Auditing taxpayers

Auditing is a necessary element of good tax administration. If information matching or cross-checking fails to identify underpayment of tax, then auditing is the only way to uncover intentional noncompliance. Typically, auditing means the examination of filed returns by tax authorities to determine the correctness of self-assessed taxes. The authorities may also use audits as the basis for statistical studies of taxpayer characteristics to be used in developing presumptive indicators— a prominent feature of taxation in many developing countries (Bird and Wallace 2004).[35] The success of auditing and the feasibility of various auditing strategies depend on the quality of the information available to the auditor, which in turn depends on three factors: the information gathered from the taxpayer and third parties, the information processing capacity of auditors, and the strategy pursued. As more advanced IT systems improve the first two factors, the authorities have a greater range of auditing strategies. In an increasing number of countries, formal systems of risk-based auditing are increasingly coming into use (Biber 2010).

In developing countries, and often in developed countries, many taxpayers play the tax lottery. A good audit strategy alters the taxpayer's perception of the odds of avoiding tax. If taxpayers are distinguishable by certain indicators (such as age, marital

---

33. See http://www.itdweb.org/documents/public/denmark.TASTSELV%20%-%20the%20auto mated%20tax%20administration.pdf
34. Interestingly, Coleman (2007) argues that this reform is unlikely to be accepted in Australia because, with the present over-withholding in that country, most Australians are accustomed to receiving a refund, and they would rather have a refund and a return than no return and no refund. One wonders.
35. For a recent survey of the special regimes to cope with hard-to-tax taxpayers in place in ten Latin American countries, see IDB (2007). See also a recent international agency study on taxing SMEs (World Bank 2008).

status, profession, or visible assets), and these indicators are correlated with reported income (i.e., are reasonable presumptive indicators), then using such indicators to refine estimates of audit results should raise the effectiveness of auditing. The Internal Revenue Service has adopted this rationale in using a statistical technique to rank taxpayers for audit selection in the United States. Even with fully computerized records and third-party information, developing and implementing such systems requires significant skills of limited availability in many developing countries. To make better use of limited administrative resources available for audit, Colombia provides taxpayers a second chance. A notice is sent to taxpayers informing them of their selection for audit and giving them the opportunity to revise their return before the audit and pay any additional taxes and interest. Those whose revised tax liability is within a predetermined percentage of the extra taxes expected of them (as calculated by a formula such as that indicated above) avoid being audited.[36]

Good case tracking of files and regular evaluation of results are essential for effective auditing. Auditors need adequate infrastructure support in the form of computer facilities, access to information which is centrally stored, quick communication facilities, and so on. When making a field visit, the audit team should have access to the full records of the taxpayer. The technology now exists for field auditors to download complete and up-to-date taxpayer files, and to check field results through secure access to encrypted mobile units. However, it takes time to use it properly, even in technologically advanced countries.[37]

[5] Services to taxpayers

Developing countries have made significant progress in using technology to improve services to taxpayers. For example, Chile reduced the burden on taxpayers by eliminating payment by cash or check, and replacing it with payment by electronic funds transfer. Singapore has gone even further, often saving taxpayers the trouble of making a payment by dipping directly into their bank accounts. Other developing countries have also taken steps to make the lives of taxpayers easier— and, not incidentally, to improve the reliability of their revenue streams. A few decades ago it was not uncommon to see long lines of taxpayers in the street in front of the local tax office when the time came to make the monthly, quarterly, or annual tax payments. Even when computers were first introduced, taxpayers frequently had to line up to get the forms that they had to submit. Such sights are now largely history in most parts of the world. Now the painful act of paying taxes can often be done from one's home computer— at least in countries in which the electricity supply is reliable and in which most taxpayers have access to computers. Neither condition holds in many developing countries, notably those in sub-Saharan Africa, but they do hold, increasingly, even in countries like India that have huge poor populations but also increasingly large middle

---

36. As Vazquez-Caro and Slemrod (2005) discuss, Colombia's tax administration remains far from ideal in many ways, including auditing practices.
37. A decade after Canada adopted a VAT, the Auditor General (1999) still found it necessary to push the revenue agency to do more and better VAT audits.

classes. Even taxpayers without a computer can usually pay their taxes at any local bank branch in many countries, and (as mentioned above) payment through mobile telephones may also become more common in some developing countries.

Different countries provide different services on the web. As a rule, such services were initially extended on a voluntary, not mandatory basis so that taxpayers did not have to file (or pay) in this way unless they perceived sufficient benefits to themselves from doing so. Judging from the 'market test' of high take-up rates of such services in countries such as Denmark and Chile, however, it seems likely that further moves to web-based tax administrative systems are the wave of the future. Countries such as Singapore, Chile, and South Africa already have excellent web pages that provide taxpayers with information about their obligations, answer many taxpayer questions, and provide return-free filing. Brazil has managed to extend electronic filing coverage to 90% of its taxpayers. Even when governments do not provide such facilities, tax preparation software is often readily available to taxpayers and is usually acceptable for official purposes.

[6]   Management of tax administration

In many developing countries, the different tax administrations concerned with internal taxes, customs, and social security, fail to share information or coordinate the processing of returns and the maintenance of records. This state of affairs clearly leads to duplication and a loss of administrative effectiveness. The effective use of new technology can help resolve these problems.[38]

Adopting new technology has permitted major changes in tax administration, structure, and operation in many developing countries.[39] However, competent accountants and careful application of already existing technology remain more essential in improving tax administration in most developing countries than bold and creative technological innovation.

Finally, properly developed technology systems have the potential to reduce corruption in part by reducing face-to-face interaction.[40] In some countries, taxpayers meet with tax officials only when the officials are soliciting bribes. Removing the

---

38. In the Philippines, database links between customs and internal revenue have been used to match VAT declarations and imports. SAS, Customer Success: Philippines Bureau of Internal Revenue, http://www.sas.com/success/philippinesbir.html. In Peru, data mining software has reportedly allowed tax authorities to reduce customs fraud and tax evasion by 14%. SAS Helps Peruvian Tax Authority Reduce Customs Fraud and Tax Evasion by 14 Percent, http://www.sas.com/news/preleases/021805/news1.html. The computerization of back-office functions in South Africa has permitted the shifting of increased personnel to front-line customer relations. Areff and Mabaso (2005).
39. For other country examples, see UN (2007) and FIAS (2006).
40. For example, in Karnataka, India, 7 million farms can now obtain printed copies of land titles (which they need several times a year to secure bank loans) online in ten minutes at government-run or privately operated Internet kiosks, for a fee of 15 rupees (about 33 cents). Under the previous titling system, two-thirds of users reportedly had to pay bribes much greater than this fee. In contrast, only 3% of users of the online system report paying bribes. Sometimes technology may increase corruption by increasing the opportunities for more sophisticated collaboration between taxpayers and corrupt officials.

opportunities for corruption (and harassment) is one reason often given for introducing various simplified, and presumptive substitute, tax systems (Engelschalk 2004). More generally, the adoption of new technology in developing countries may facilitate the drastic change that is needed in tax authorities' attitude and operations if countries are to move from treating taxpayers as thieves to be caught to viewing them instead as (reluctant) clients to be served.[41]

§4.03 TECHNOLOGY AND THE DESIGN OF TAX SYSTEMS

Our discussion of technology and tax administration does not lead to any radical new conclusions, but then for the most part we were covering well-trodden ground. In contrast, how technology may influence the design of tax systems or the design of specific tax instruments in developing countries is a much less discussed question. We first consider some general questions about tax system design in developing countries, and then look briefly at some proposals with respect to specific tax instruments that may be more feasible as a result of improvements in technology.

[A] Tax System Design

Technological advances will alter the economic environment in which governments seek to collect tax revenue. These advances will make some persons or transactions easier to tax. Particularly in developing countries, use of various methods of electronic payments will move more transactions from the informal to the formal economy. As discussed earlier, technology will also provide tax administrators with more tools to track the movements of goods and individuals. It should also increase opportunities and reduce costs of cooperating with tax administrators in other countries to improve tax compliance of persons with investments and activities outside the country.

But technology will also make some persons or transactions harder to tax. For example, it is currently much harder for tax authorities to track goods in digitized form than those that are physically transported across or between countries. Foreign lawyers, accountants, and management consultants can provide services with little or no physical presence in a country. Advances in the financial service industry allow domestic investors access to foreign banks and securities with simple Internet access. The globalization of financial markets has made it harder for any one country to tax income from mobile capital.

Technology may also influence incentives of governmental officials in the design and administration of tax systems. Hettich and Winer (1999) set out a model in which changes in administrative costs may affect both the size of the public sector and the choice of tax structure. Costs may change because of changes in administrative technology like those we have just discussed, or they may change because technology

---

41. The importance of such a shift in attitude is emphasized in Bird (2004): for some of the reasons, see, for different but complementary views, Moore (2004) and Kirchner, Hoelzl and Wahl (2008).

has changed the nature of the economy.[42] For example, as we noted earlier, the more developed the financial structure of a country is, the more likely it is to rely more heavily on 'modern' taxes such as the VAT (Gordon and Li 2009).

Technology may have a double-edged impact on tax complexity. On one hand, the greater flexibility afforded tax policy designers allows them to introduce ever more refined classifications and categorizations into legislation— for example, targeted client-specific provisions (e.g. low-income allowances) can more readily be implemented at lower administrative and compliance cost. On the other hand, techniques such as pre-populated returns and web-based returns (e.g. keyed to plain-language explanations of terms) may make it simpler for taxpayers to comply even with complex laws.

Indeed, one important lesson emerging from experience in various countries is that an essential precondition for the reform of tax administration is sometimes to simplify the tax system to ensure that it can be applied effectively in the generally low-compliance contexts of developing and transitional countries.[43] Tax reforms in both Chile and Colombia demonstrate that considerable improvements can be made in administration with less drastic, but nonetheless effective, simplifications in tax policy. Reducing the number of income tax deductions, for instance, permitted these countries to eliminate filing requirements for most wage earners, thus greatly reducing the administrative burden as most income taxpayers were able to fulfill their obligations solely through withholding.[44]

Tax systems vary greatly in their reliance on different tax instruments to generate revenue in order to support government operations. Bird and Zolt (2005) examine how developing countries rely much more on trade taxes, excise taxes, and general consumption taxes, and less on income tax, than developed countries. The relative use of different tax instruments is influenced by many factors, including economy-specific characteristics, the ability of taxpayers to understand, comply with, or evade, tax provisions, and the ability of taxing authorities to administer the tax laws. In addition, it is partly explained by the different costs of different tax technologies in developed countries, as compared to that in developing countries.

Technology may change this calculus of tax structure. First, as we discussed above, it may enable tax authorities to change the policy mix towards taxes that can now be more effectively administered. Secondly, the characteristics of an economy influence the design of a tax system. As discussed earlier, the classic developing country profile consists of a large subsistence agricultural sector, a substantial small

---

42. As mentioned earlier, almost nothing is known about the relevant cost curves. However, for an interesting econometric exploration of this hypothesis using aggregate country data, see Kenny and Winer (2007).
43. The experience of Bolivia, which introduced a major simplification of its tax system in 1986, is instructive. Much of the initial success achieved in reforming Bolivian tax administration was clearly attributable to the extensive simplifications made in the tax system (Silvani and Radano 1992). Indeed, as Bahl and Martinez-Vazquez (1992) argue in the case of Jamaica, it seldom makes sense to reform tax administration without simultaneously reforming tax structure.
44. For Chile's experience, see Harberger (1989). For Colombia's experience, see McLure and Pardo (1992). Some countries go the other way, as shown in the discussion of the Polish VAT in Bird (1999).

service sector marked by cash transactions, small retail establishments, and the relative absence of large employers (other than the government and a few domestic and foreign companies). Technology may change both the size and structure of this tax base.

Technological change may, for example, reduce the threshold for effective taxation, whether based on the size of a transaction or the size of a firm. Developed countries are more effective than developing countries at taxing consumption and income in large part because taxpayers have more adequate books and records and tax authorities have greater ability to observe and monitor transactions. The introduction of electronic cash and smart cash registers has in principle dramatically reduced the size of the transaction or firm that authorities can observe. As such technologies spread to serve the expanding middle class in many developing countries, the capacity of tax administrations to observe transactions will improve.[45]

By changing the relative compliance and enforcement costs and compliance rates of different tax instruments, technology influences the relative revenue that can be generated by different taxes. Technology may not only allow policy-makers to change thresholds in such a way as to increase the scope of some taxes. It may also permit decreases in regressivity or increases in progressivity of specific tax instruments that may allow them to be more significant revenue contributors. More than that, by lowering economies of scale in tax administration, improved technology may increase the feasibility of collecting income or value-added taxes at sub-national levels of government. Decentralizing important revenue sources might be one way to increase the effectiveness of government spending (Litvack et al. 1998). Technology may also foster improved measurement of the benefits provided to specific individuals or groups by government spending and hence increase opportunities for benefit taxation. Linking revenues and expenditures in such visible and meaningful ways might be an important way for the tax system to contribute to the building of social capital in developing countries.[46]

[B] Tax Instruments

[1] Value-added taxes

The VAT has emerged as the major source of tax revenue in developing countries (Keen and Lockwood 2006). In sub-Saharan Africa, the VAT is about 30% of total tax revenue (Cnossen 2006a). In Asia, the VAT contributes about 21% (Cnossen 2006b). Many major design issues in the VAT are affected by changes in technology. One issue that comes up in many countries, for example, is the coordination of VAT information with information from the customs authority. In Ghana, the Ghana Community Network (GCNet) was created to modernize Ghana's customs operations. For instance, shipping lines provide electronic manifests to GCNet, which are then transferred to the Ghana

---

45. Tax evaders may also have access to advanced technology. For an interesting account of the technological fight against fraud in Quebec, Canada, see Ainsworth and Hengartner (2009).
46. For interesting reflections on this topic, see Brautigam, Fjeldstadt, and Moore (2008).

Ports and Harbors Authority, the Ghana Shipping Council (which obtains all information regarding the movement of ships and airplanes), customs (which obtains customs goods declarations electronically), banks (which inform customs electronically of payments made), and the Ministry of Finance (which can download all trade information as well as all transactions of taxpayers identified by personal identification number) (De Wulf and Sokol 2003).

As noted earlier, the use of electronic invoices increases the potential for matching invoices. Not only does this allow tax authorities to match identifications of buyers and sellers, but if VAT taxpayers are permitted to access the taxpayer registry, they can check that they are in fact buying or selling to legitimate VAT taxpayers. Long before technology made it feasible, Mushkin (1972, 407) presciently suggested that the regressivity of the VAT (or other consumption taxes) could be reduced by providing differential treatment to low-income taxpayers (Ainsworth 2008). The use of smart cards would permit reducing taxes on food and other items only for low income taxpayers, perhaps making it even more likely that developing countries will continue to rely heavily on consumption taxation.

[2] Excises

In many developing countries, excise taxes currently raise more revenue than individual income taxes. On average, excise taxes are about 19% of tax revenue in sub-Saharan Africa and 16% in Asia. The primary products subject to excise taxes are tobacco, alcohol, gasoline, and motor vehicles. New technology would, for example, allow countries to adopt differential rates for high and low value alcohol and tobacco (using encoded information), or to reduce rates for targeted populations.

[3] Trade taxes

As with excises, trade taxes are much more significant in developing countries than developed countries. Although trade liberalization and the attraction of joining the World Trade Organization has significantly reduced trade tax revenue in developing countries, trade taxes still constitute about 25% of total tax revenues in low-income countries, compared to less than 3% in high-income countries (Fox and Gurley 2005).[47] Generally, taxes on imports are of much greater importance than taxes on exports. Curiously, this oldest form of taxation likely has been the object of the greatest technological innovations and applications as taxing authorities around the world seek to improve their ability to monitor and value goods entering the country. Three examples of such technology are Tamper-resistant Embedded Controllers (TRECs), Radio-frequency ID (RFIDs), and Laser Surface Authentication (LSA). The EU has a program using TRECs to monitor container shipments with greater security and

---

47. As Keen (2007) shows, in addition, much VAT in many developing countries is also collected at the border, and much of the VAT collected at import is not subsequently credited against output VAT on later sales. In effect, it is a final tax imposed on informal economic activities.

efficiency.[48] If the container deviates from the pre-programmed shipping route, the system notifies the interested parties (Swedberg 2006).[49]

TRECs are fairly expensive and bulky and economically feasible only for large shipping containers. But RFID tags are relatively small and inexpensive and can be used to track individual packages. Such tags are currently used by retailers such as Wal-Mart to track shipments from suppliers, and to monitor inventory, as well as by pharmaceutical companies to reduce the risk of counterfeit drugs entering their supply chain.[50] An even more advanced technology is LSA, which generates a unique digital serial code for an item by analyzing the surface of that item. This code acts as an individualized fingerprint for each item and can thus be used to identify each item uniquely. Since the code naturally occurs from microscopic imperfections that exist in the manufacturing process, products can be tracked without the addition of any chips, inks, or other tags ("Customs and Biometrics" 2006). Smugglers may be getting smarter, but so are the systems that countries use to deal with them.

[4]   Property taxes

Property taxes play a relatively small role in most developing countries (Bahl and Martinez-Vazquez 2007). Revenues from property taxes generally account for little more than a 0.5% of gross domestic product (GDP). One reason for their relatively small role may be political, as those who control or influence the political process are often those who would be disproportionately burdened by property taxes.[51] In many developing countries, the lack of reliable surveys and records, and the difficulty in valuing land and improvements also limit the use of property taxes. While political constraints will remain important, improvements in technology make it easier to have an effective property tax system if a country really wants to do so. Given the prevalence of decentralization around the world, taxes that can be effectively implemented at a local level (such as property taxes) may become more important.

Effective property taxation requires adequate records to identify property and locate owners, and a reliable method of valuing property. Internet-based resources

---

48. Each Tamper-resistant Embedded Controller (TREC) contains a microprocessor, a Global Positioning System (GPS) receiver, and links to various sensors on the containers. Using its GPS device, the TREC wirelessly transmits the location of each container. For more information on IBM's TREC platform and its Secure Trade Lane Container Information Services, see Dolivio (2007).
49. The TREC can also sense if the container has been opened or closed, and can detect when the container has been loaded or unloaded from a vessel. In addition, RFID readers can be connected to the TREC unit to allow shippers to track RFID-tagged items inside the container.
50. Information Technology for Adoption and Intelligent Design for e-Government (ITAIDE) recently demonstrated, through a pilot project, how technology could improve the administration of excise taxes on imports (ITAIDE 2007). For information on an alternative technology that can be used to track large metal shipping containers, see Monarch (2007).
51. A particularly egregious case occurred in one small developing country in which a proposed property tax reform in the capital city was derailed when it became clear that one of the principal owners of property in that city was the wife of the country's political leader. In contrast, in another country, examination of individual property tax records revealed that, in practice, taxes were being applied only to members of a minority ethnic group.

(such as Google Earth or Microsoft's VirtualEarth service) allow for satellite imagery that assists in identifying and monitoring specific property locations and improvements.[52] In some respects, this is just a continuation of existing practices: extensive use was made of aerial photographs, for example, in establishing the cadastre (land registry) in such Latin American countries as Chile and Colombia over 40 years ago. Taxing authorities in some countries are now relying on satellite images to determine improvements in property in lieu of the corruption-prone alternatives of depending on information from other sources and walking the streets.[53] Improvements in Internet capability both increase the availability and reduce the cost of the new technology.

Internet-based valuation programs may also prove useful in at least some countries. One model, used by such sites as Zillow.com and Redfin.com, may provide taxing authorities additional information in determining the value of properties. These programs rely on publicly available data on prior sales and then use an algorithm to estimate the value of other properties. In countries where an active real estate market exists, these programs may supplement existing valuation methods.[54] In rural areas, a simplified property tax system could sort land into three or four categories based on productivity or other characteristics.[55] Here, programs such as satellite imagery that permit some assessment of soil quality may facilitate imposing property taxes.

[5] Income taxes

Finally, consider income taxes. Technology obviously offers many opportunities to alter the design of such taxes to make them simpler, more enforceable, and, if desired, more progressive. Individual income taxes may be readily coordinated with payroll or other employee taxes. Smart cards with employee-specific information may be used for purposes such as targeted credits and exemptions. If everyone who provides services is required to furnish an ID number or smart card, it would be possible to improve information reporting and withholding on payments for services and small suppliers. Businesses could claim deductions only if they provide the ID information or proof of withholding. Similar treatment could be applied to interest payments, with deductions allowed only if it was clearly demonstrated that the recipient of the payment was filing taxes properly with respect to that income.[56] It would also be much easier for taxing

---

52. For an account of recent actions along these lines in Buenos Aires, see Besfamille and Olmost (2010).
53. As often is true, of course, much the same could be done with earlier, cruder technology such as aerial photogrammetry: in one South American city, for example, examination of recent aerial photography found that buildings covered at least 45% of the land recorded in the tax roll as 'vacant'.
54. However, as experience in even the most developed jurisdictions shows, computerized methods cannot and should not completely replace traditional valuation methods (Slack, Tassonyi, and Bird 2007).
55. For a recent proposal along these lines for India, see Rajaraman (2004).
56. Alternately, countries could follow the Singapore approach for dividends, and require interest payments and dividend payments to be paid from a central fund. The government could then withhold tax at the appropriate marginal rate for every taxpayer. We do not discuss here the more complex issues raised by cross-border payments.

authorities to monitor different types of activities carried out by small businesses, for example, by tracing the flow of inputs or outputs or by video surveillance of customers' activities.[57] Improved monitoring techniques may allow tax authorities either to eliminate presumptive taxes completely (as all potential taxpayers could be included in the normal tax base), or to make such taxes more effective and easier to integrate with the regular tax system.[58]

[C] Tax Reform Process

Good tax administration relies on the good use of information. Good tax reform similarly depends on the good use of information. Improvements in information technology allow countries to administer existing taxes better, and to change both the administration and the structure of taxes to better achieve their developmental objectives. Techniques for developing tax forecasting and tax simulation models are now advanced enough— which, paradoxically, means they are simple enough to be understood and used by anyone at the university level— that even small and unsophisticated tax departments can use such models to provide information to policymakers. Better information about predicting consequences of tax reform proposals, improved ability to determine the revenue and distributional consequences of proposed tax changes, and better ability to forecast future government tax receipts, should lead, over time, to better designed and administered tax systems in developing countries.

§4.04 PRIVACY AND PRIVATIZATION— SOME CAUTIONARY NOTES

In developing countries, as in developed countries, the adoption of new technology carries with it potential pitfalls as well as potential gains. Here we discuss briefly two areas with potential problems: privacy and privatization.

[A] Privacy— Big Brother Is Watching

Schwartz (2008) provides a general overview of privacy issues relating to tax and technology. Privacy scholarship has focused primarily on issues in developed countries. Here, we examine privacy issues in the context of developing countries.

Conceptions of rights to privacy differ among societies.[59] It is difficult to determine the scope of privacy rights without a particular context, whether it is risks of

---

57. Of course, a good field audit can do the same: for an example of excellent field audit skills, see the activities of Ryoko Itakura, the tax inspector in Juzo Itami's movie, *A Taxing Woman* (1988).
58. For a detailed discussion of the limitations of most existing presumptive income tax systems, see Bird and Wallace (2004).
59. Europeans and Americans apparently have different conceptions of privacy rights. Europeans focus more on a dignity-based concept, whereby privacy is violated when there is an unauthorized portrayal of the individual. In contrast, Americans focus more on a liberty-based approach, whereby privacy is violated when the state makes an unauthorized intrusion on the rights of the individual (Whitman 2004).

tax evasion or risks of terrorism. Our focus here is on the use (and abuse) of information by tax authorities. The challenges and scope of privacy rights extend beyond tax to the use of information by other government agencies, other governments, or private actors. Improved technology and access to information has made it possible for anyone to obtain the salaries of professors at public universities,[60] the political contributions of friends and foes,[61] and the value of the houses of one's neighbors.[62]

Improvements in technology have also made it easier for tax administrators to observe transactions and taxpayers. In the last decade, there has been a substantial increase in the ability to trace transactions (the tracing of payments and the tracking of physical products), the use of smart identification cards to match individuals to specific jobs and locations, and the use of devices (such as photo surveillance and traceable mobile telephone technology) to monitor movements of individuals.[63] The final outcome of the advent of the digital economy may thus be to strengthen, not weaken, the government's role as tax collector. At the extreme, as Rädler (2000, 798) put it, "[I]n a few years each of us will know his or her uniform global tax identification number to be used on a world-wide basis. It will give our place of residence ... cash will totally disappear ... its use will fade away because everybody will be obligated to exclusively use his money-card."[64] In Rädler's perhaps dark vision of the world, "Income taxation may be rather simple: most of the information needed is already contained in the central computer and will be correspondingly processed. The taxpayer has only to check whether the information provided by the computer is correct." This is not simply a utopian (or distopian) vision. To some extent it is already a reality in Singapore, where withholding has been taken to its logical limit so that the government can transfer funds directly from a person's bank account to the treasury to settle tax liabilities as calculated by the government.

Some have suggested that the real danger from new technology may not be the erosion of the tax base as taxpayers use technology to avoid tax, but the erosion of privacy as governments take defensive action to protect the fisc in the emerging digital economy. Perhaps the only viable answer is for citizens (at least in countries in which they have a say in such matters) to reach a consensus that permits access to private activities necessary for the sustenance of the public community, without allowing such information to be misused. One author's answer to this dilemma was what he called "the transparent society"— one in which, above all, people are held accountable for their actions, including what they do with the information to which they have access (Brin 1998).[65] In other words, although presumably none of us wants our foibles and

---

60. http://www.sfgate.com/news/special/pages/2005/ucsalary.
61. http://www.opensecrets.org.
62. http://www.zillow.com.
63. For example, in recent years divorce lawyers have made increasing use of information from toll tags to establish the existence of extra-marital relationships. http://www.msnbc.msn.com/id/20216302.
64. See also "Barter's Latest Comeback," *The Economist* 21 (October 2000): 78.
65. In a comment on this possible future, Manasian (2003, 25) suggested, rather despairingly perhaps, that the "antidote to the rapid erosion of personal privacy may prove to be not new

weaknesses publicly available, more transparency may simply be yet another part of the price we pay for living in a complex modern society.

Countries around the world, developed or developing, need to balance the benefits of providing taxing authorities with additional tools and resources, against the costs of invasion of privacy and potential for abuse by government officials and others. (Swire 1999). Different countries may strike different balances. Given the high levels of tax avoidance and the large size of the informal economy in most developing countries, it is likely that the benefits from providing these technological tools and resources to tax officials are greater in developing countries than in developed countries. Unfortunately, as discussed below, the potential for abuse and the associated costs are also likely to be greater in developing countries.

The potential gains to tax authorities from increased use of information arising from improvements in technology are clear. However, the availability of different types of information may also result in government officials using information for financial gain, political gain or discrimination, or simply for the thrill of invading the privacy of well-known individuals. Government officials could also use information, not for personal gain, but in a manner inconsistent with the laws governing privacy in order to achieve what they believe is good government policy. In addition, the potential exists for unauthorized third parties to hack into government computers for their own financial gain or political purposes.

Bennett (1991) frames the privacy-protection approach nicely by focusing on technology control, civil rights, and institutional accountability. Tax authorities can adopt safeguards for the different stages of information processing: input, storage, and output. Such safeguards include password protection, encryption, proper training of computer staff, and otherwise limiting access to the information. Swire (1999, 477-85) uses the analogy of the vault 600 feet down as a device in order to illustrate how societies could make choices about how to access information, how to limit unauthorized uses, and how to limit the disclosure of information with high privacy risks.

To date, most privacy legislation in developed countries has focused on protecting the rights of individuals. For example, the EU Data Protection Directive provides rules for private-sector processing of personal information. The US Privacy Act of 1974 sets forth guidance on the use by the Federal government of data about individuals. Section 6103 of the Internal Revenue Code provides rules on the disclosure of tax return information.[66] The US Treasury has also issued regulations covering the disclosure or use by tax return preparers of tax return information.[67]

A major challenge is to design regimes in order to ensure institutional accountability. There is nothing unique about increased use of technology in tax design and administration. In both developed and developing countries, the key issue is whether the legal and political institutions provide sufficient checks and balances to allow use

---

laws but new rules of etiquette." Dependence on the politeness of tax collectors is not likely to strike most citizens as a very firm reed on which to rely.

66. For access to records from the Internal Revenue Service, see generally Saltzman (1991, paragraph 205).
67. http://www.irs.gov/irb/2006-03_IRB/ar16.html.

of the information generated by improvements in technology without the likelihood of substantial abuses.

Given the diversity of cultures and societal norms, there likely exists a wide range of tolerance and intolerance for measures that infringe on privacy in different developing countries. Privacy International, a human rights group, surveys developments in 70 countries, assessing the state of technology, surveillance, and privacy protection.[68] Many policy initiatives adopted by governments may challenge personal privacy. Such initiatives include many of the technologies that may be used to improve the tax system discussed earlier: the use of identity cards using fingerprint and iris scanning biometrics, the linkage of public sector computer systems, the development of real-time tracking and monitoring through the communications spectrum, the development of real-time geographic vehicle and mobile telephone tracking,[69] national DNA databases, and the creation of global information sharing agreements.

Privacy International ranks countries according to constitutional protection, statutory protection, privacy enforcement, identity cards and biometrics, data sharing provisions, visual surveillance, communications interception, workplace monitoring, law enforcement access to data, communication data retention, surveillance of travel and finances, and global leadership and democratic safeguards. In the national privacy rankings for 2006, the countries with the lowest rankings include not only such expected candidates as Russia and China, but also Singapore and Malaysia— two countries that would likely rank near the top of most lists of developing countries with respect to the quality of their tax administration.[70] Developing countries with membership in the EU score relatively well, as do several Latin American countries. Although not rated, we suspect that Chile, which also ranks high at least on our list of developing countries with good tax administrations, would also score well in privacy rankings. On the whole, however, it seems probable that if tax administrators were to improve significantly their capacity to acquire information, few developing countries would have adequate procedural safeguards to offer sufficient privacy protection— at least from the perspective of current standards in most developed countries.

The costs of disclosure of information may also differ among developing countries. In Colombia, for example, one reason offered by some for the recent discontinuation of the long-standing wealth tax (in existence since 1935) was fear that the misuse of this tax information might increase the risk of kidnapping.[71] In environments in which lives may be at stake when tax information is misused, the barriers against such misuse should clearly be very high. When it is not possible to erect and enforce adequately such barriers, it may perhaps make sense not to collect the information in

---

68. http://www.privacyinternational.org.
69. Some Brazilian states employ GPS tracking devices backed up by mobile units of tax officials to keep track of cross-border trucking. The Brazilian state VAT is essentially levied on an origin basis which requires some check at the state border. Illustrating how good technology can help overcome poor tax design, appropriate use of IT obviously facilitates this archaic process.
70. The UK also ranks near the bottom of the list because of its widespread use of surveillance cameras. If concerns about personal security continue to trump concerns about privacy in other developed countries, they too may fall in the privacy rankings.
71. Cynics may suspect that this argument hides less worthy motives. That may be, but the fear is nonetheless real.

the first place, even if many of the nicer distinctions among persons and the tax base that one might like to see in place in a good tax system suffer as a consequence.

[B] Privatizing Tax Administration

Many technology-based proposals to change tax administration raise questions about the role of the private sector in the assessment and collection of taxes. This public-private issue for tax regimes is not new to issues of technology. The allocation of tasks to private actors has a long tradition in tax systems. The Romans, Egyptians, Ottomans, Mamluks, and the French, each sold rights to collect taxes in geographic regions in return for a fixed fee. Other states have used privatized tax collection, whereby private actors collect taxes for the state in return for a percentage of the take.[72]

A large academic literature exists examining the advantages and concerns arising from privatization of services that were traditionally provided by the government. Privatization covers a wide range of arrangements, from outsourcing certain limited tasks to assigning full responsibilities of certain formerly government activities to private actors. Examples in developing countries range from the use of pre-shipment inspection to establish valuations for customs purposes in countries from Indonesia to Liberia, to the out-sourcing of customs administration in Mozambique to a private consulting firm.

Those who support increasing the role of the private sector in providing government services, note that several advantages might result from private initiatives. These advantages include the potential for improvements in the quality and effectiveness of services, increased competition and incentives for innovation, and the ability to avoid limitations that would be tied to government employees providing the services directly. (Trebilcock and Iacobucci 2003; Hart, Schleifer, and Vishny 1997). Concerns include the potential mismatch between private profit incentives and the social provision of goods, the loss of accountability for private actors, and the loss of public support for public institutions (Minow 2003). This last factor is especially important in the privatization of tax functions in countries where support for tax systems and tax administrators is weak.

While this is not the place to review the competing considerations of private production and government provision of services we agree with the approach that in examining the merits of outsourcing or privatization, the analysis must be relative: the relative advantages of using the private sector over government employees, and the relative potential (and costs) for private sector failures as compared to government failures (Trebilcock and Iacobucci 2003). Significant spillover potential exists from adapting private sector applications to improve government operations. Whether using technology (originally designed for inventory management) for identification and valuation purposes by customs officials, or using data management programs to allow sharing of information among different government agencies, private sector expertise

---

72. The recent use by the Internal Revenue Service of private debt collection agencies highlights the advantages and concerns of using the private sector in tasks that have been traditionally performed by government employees.

and services clearly provide great potential benefits for tax design and tax administration. The question is how best to achieve these benefits without incurring unnecessary social costs.

The key challenge in outsourcing technology-based services is the loss of control of the confidentiality of taxpayer information, the protection of taxpayers' rights, and the tax administration's ability to have the expertise to continue to perform their activities. Again, the comparison is between the potential for failure for private sector activities against the potential for failure if the government provided such services directly, and the costs of such failures.[73] Even such simple cases as the use of private banks to receive and process returns from taxpayers carries with it potential problems as well as benefits. Just as oversight and process controls are needed to monitor the actions of government employees, similar measures are required to monitor the private provision of historically government functions. In countries where administrative, judicial, and political safeguards work relatively poorly when overseeing government employees, there is little reason to believe that such safeguards would work better when monitoring the actions of private actors.

The outsourcing of technology-based services in developing countries raises additional challenges. First, where foreign donors provide financing for such services, there may be direct or indirect pressure to select certain vendors or suppliers who may or may not be the best provider. Second, even where no outside pressure exists, developing countries may lack the expertise to select and monitor outside contractors. Finally, where whole functions or departments are outsourced, no internal capacity may be developed, and there will be challenges in taking over the tasks at the expiration of the contracts.

§4.05 THE FUTURE RELATIONSHIP OF TAX AND TECHNOLOGY

Our review of tax and technology leads us to the following four conclusions. First, improvements in technology are not magic bullets to improve tax policy or tax administration in developing countries. Over the last 40 years there have been significant technological advances, but developing countries still have relatively low levels of tax collections as a percent of GDP, and relatively high levels of tax non-compliance. The challenges to designing and implementing effective tax regimes in these countries cannot be met solely by better tools for tax administrators, without substantial changes in the institutional and political environment. Building a good tax system is much like building a good state. In fact, both activities are likely to advance more quickly and soundly if they are undertaken together (Brautigam, Fjeldstad, and Moore, 2008).

---

73. An historic example is the Chinese Maritime Customs imposed on China as a means of servicing foreign debts in the late 19th century. For our purposes, the question is neither the efficacy of this foreign controlled agency in collecting the required revenues, nor the fact that some of the foreigners involved in the agency ended up with very large personal fortunes. The key question is whether tax administration in China was, on balance, benefited or damaged by this compulsory outsourcing of a principal revenue function.

Second, technological innovation makes it essential to reexamine tax policy and administration in developing countries. Many low-income developing countries still rely on the basic tax regime inherited from colonial powers.[74] Most tax revenues come from trade taxes and other taxes on consumption such as excise taxes and value-added taxes. Technological innovations has and will continue to change the way individuals and firms operate in the economy in ways that will make it both easier and harder for tax authorities to observe and monitor their transactions. These changes will give rise to additional choices for developing countries in choosing the relative role of different types of taxes in raising revenue and in designing the tax instruments.

Third, the potential gains from improvements in technology are likely greater in developing countries than developed countries. This assertion rests partly on the existing high levels of non-compliance, and the presence of more low-hanging fruit in developing countries relative to developed countries. Most developed countries have already adopted information reporting systems that cover a substantial portion of economic activity. Improving the capacity of tax authorities in developing countries to observe and monitor transactions will yield substantial benefits. Technological improvements reduce compliance costs and delays associated with paying taxes (e.g., clearing goods at customs or standing in line to pay taxes), and such costs are generally higher in developing countries. Technology will also provide tax administrators with better tools to tax more effectively the informal economy and informal employment arrangements.

Fourth, the potential costs of using more sophisticated technology may also be greater in developing countries than in developed countries. Privacy concerns exist in all countries. Countries have adopted different approaches to minimize the loss of privacy rights from government or business activity. Although some developing countries have been successful in adopting administrative, judicial, and political safeguards (such as Argentina), most developing countries lack effective mechanisms to help prevent or discourage the abuse of information by government officials, or to limit access by non-governmental entities.[75]

New technology may result in the increased use of public-private partnerships, whereby private firms will take responsibility for tasks usually handled by government employees. The advantages and disadvantages of this type of privatization are similar in both developed and developing countries. What may differ, however, is the likelihood of introducing adequate institutional safeguards to minimize the potential costs from allowing private firms to assist with, or to take over, tax functions of the government.

Given the diversity among developing countries, it is difficult to make generalizations as to the future relationship between tax and technology. Some economies in developing countries are growing at over 10% a year with a high rate of technological

---

74. To take an extreme example, until recently the basic income tax nominally in force in Iraq was that imposed under the British mandate in the 1920s.
75. Developed countries also face substantial challenges in preventing unauthorized access to government records. In the United Kingdom, at one point the authorities lost detailed tax identification and financial information on 25 million taxpayers.

change. Other developing countries (or, more accurately, non-developing countries) have stagnant or declining growth rates, and even declining life expectancies in a few instances. With this qualification, we offer the following observations.

First, like developed countries, the higher-income developing countries will continue to experience dramatic changes in their economy from technological and other factors. The increased use of banking systems, credit cards, smart cards, and online transactions will present both opportunities and challenges to tax authorities. This new economic environment will allow for improvements in tax compliance for some taxes and more challenges for other taxes.

Second, further improvements in tracking technology for both people and goods will provide additional tools for tax authorities to improve tax compliance. With its declining costs and improved capability, the use of tracking technology by tax administrators in developing countries will increase. For example, it is likely that many capital goods (cars, heavy machinery, or even televisions) will have identification and tracking tags included as part of the manufacturing process. This ability to monitor the movement of goods will provide new opportunities in customs, excise, and VAT administration.

Third, perceptions of increases in crime and terrorism will likely lead to higher levels of government surveillance and often greater acceptance of government monitoring of individual and business activity. For example, developed countries have made greater use of video cameras, and of tracking funds through banking channels. Combining video technology with smart identity cards will allow officials to track the movement of individuals. These and other technological advances will make it easier for government officials to monitor economic activity. As Slemrod (2006) noted, however, it does not follow that tax authorities will (or should) have access to such information.[76]

In conclusion, advances in technology will clearly change the tax environment in developing countries by changing the underlying economy. Technology will provide additional tools to tax administrators to observe and monitor individuals and transactions. This combination will provide an opportunity for countries to make tax policy changes both as to the relative role of different taxes in financing government, and as to the design of specific tax instruments. With potential benefits come substantial potential costs. Most notably, developing countries need political, administrative, and judicial safeguards to protect the privacy of individuals and to protect against potential misuse of information gathered for tax or other purposes.

§4.06 REFERENCES

Ainsworth, Richard T. "Biometrics: Solving the Regressivity of VATs and RSTs with 'Smart Card' Technology." *Florida Tax Review* 7 No. 10 (2006): 651–84.

---

76. See Cockfield (2007) for discussion of some of the privacy issues arising from new state surveillance techniques.

Ainsworth, Richard T. "A Technological Approach to Reforming Japan's Consumption Tax." *Tax Notes International* 8 No. 49 (February 4, 2005): 437–47.

Ainsworth, Richard T. & Urs Hengartner. "Quebec's Sales Recording Module: Fighting the Zapper, Phantomware, and Tax Fraud with Technology," *Canadian Tax Journal* 57 No. 4 (2009) 715-61.

Alm, James, Jorge Martínez-Vazquez & Sally Wallace. *Taxing the Hard–to–Tax: Lessons from Theory and Practice.* Amsterdam: Elsevier, 2004.

Alm, James, Friedrich Schneider & Jorge Martinez-Vazquez. "'Sizing' the Problem of the Hard–to–Tax." In *Taxing the Hard–to–Tax: Lessons from Theory and Practice*, edited by James Alm, Jorge Martinez–Vazquez & Sally Wallace, 12–55. Amsterdam: Elsevier, 2004.

Areff, Sameer & Thandi Mabaso. "Maximizing Responsiveness at the South African Revenue Service." In *Revenue Matters: A Guide to Achieving High Performance Under Taxing Circumstances*, edited by Brian Dustrud & Lisa Barr, 2005. Available at http://www.revenueproject.com/ documents.asp?d_ID=3286.

Auditor–General. *Report of the Auditor General of Canada.* Ottawa: Auditor-General, 1999.

Baer, Katharine, Olivier P. Benon & Juan A. Toro Rivera. *Improving Large Taxpayers' Compliance: A Review of Country Experience.* Occasional Paper No. 215. Washington, D.C.: International Monetary Fund, 2002.

Bahl, Roy W. & Richard M. Bird. "Tax Policy in Developing Countries: Looking Back— and Forward." *National Tax Journal* 61 No. 2 (June, 2008): 279–301.

Bahl, Roy W. & Jorge Martínez–Vazquez. "The Property Tax in Developing Countries: Current Practice and Prospects." LILP Working Paper. Cambridge, MA: Lincoln Institute of Land Policy, 2007.

Bahl, Roy W. & Jorge Martínez–Vazquez. "The Nexus of Tax Administration and Tax Policy in Jamaica and Guatemala." In *Improving Tax Administration in Developing Countries*, edited by Richard M. Bird & Milka Casanegra de Jantscher, 66–110. Washington, D.C.: International Monetary Fund, 1992.

Bennett, Colin J. "Computers, Personal Data, and Theories of Technology: Comparative Approaches to Privacy Protection in the 1990s." *Science, Technology, & Human Values* 16 No. 1 (Winter, 1991): 51–69.

Bennett, John & Saul Estrin. "Informality as a Stepping Stone: Entrepreneurial Entry in a Developing Economy." Centre for Economic Development and Institutions Working Paper No. 07–11. West London: Brunel University, 2007.

Besfamille, Martin & Pablo Olmos. "Inspectors or Google Earth? Optimal Fiscal Policy under Uncertain Detection of Evaders. Paper presented at 2010 IRS Research Conference, Washington, DC June 2010.

Biber, Edmund. *Revenue Administration: Taxpayer Audit— Development of Effective Plans.* Technical Notes and Manuals. Washington D.C.: International Monetary Fund, 2010.

Bird, Richard M. "Tax Policy and Tax Administration in Transitional Countries." In *International Studies in Taxation Law and Economics*, edited by Gunner Lindencrona, Sven-Olof Lodin & Bertil Wiman, 59–75. London: Kluwer Law International, 1999.

Bird, Richard M. "Administrative Dimensions of Tax Reform." *Asia–Pacific Tax Bulletin* 10 No. 3 (2004): 134–50.

Bird, Richard M. "Taxing Electronic Commerce: The End of the Beginning?" *Bulletin for International Fiscal Documentation* 59 No. 4 (2005): 130–40.

Bird, Richard M. & Milka Casanegra de Jantscher, eds. *Improving Tax Administration in Developing Countries*. Washington, D.C.: International Monetary Fund, 1992.

Bird, Richard M. & Pierre-Pascal Gendron. *The VAT in Developing and Transitional Countries*. New York: Cambridge University Press, 2007.

Bird, Richard M. & Oliver Oldman. *Improving Taxpayer Service and Facilitating Compliance in Singapore*. PREM Note No. 48. Washington, D.C.: The World Bank, 2000.

Bird, Richard M. & Sally Wallace. "Is it Really so Hard to Tax the Hard-to-Tax? The Context and Role of Presumptive Taxes." In *Taxing the Hard-to-Tax: Lessons from Theory and Practice*, edited by James Alm, Jorge Martinez-Vazquez & Sally Wallace, 121–58. Amsterdam: Elsevier, 2004.

Bird, Richard M. & Eric M. Zolt. "Redistribution via Taxation: The Limited Role of the Personal Income Tax in Developing Countries." *UCLA Law Review* 52 No. 6 (August, 2005): 1627–95.

Bradsher, Keith. "China Enacting High-Tech Plan to Track People." *New York Times* (August 12, 2007): A1.

Brautigam, Deborah, Olaf Fjeldstad & Mick Moore, eds. *Taxation and State-building in Developing Countries: Capacity and Consent*. New York: Cambridge University Press, 2008.

Brin, David. *The Transparent Society*. Reading, MA: Perseus Books, 1998.

Casanegra de Jantscher, Milka. "Administering the VAT." In *Value Added Taxation in Developing Countries*, edited by Malcom Gillis, Carl S. Shoup & Gerardo P. Sicat. A World Bank Symposium. Washington: World Bank, 1990.

Chen, Martha A. "Rethinking the Informal Economy: Linkages with the Formal Economy and the Formal Regulatory Environment." Department of Economic and Social Affairs Working Paper No. 46. New York: United Nations, 2007.

Choi, Kwang. "Value-added Taxation: Experiences and Lessons of Korea." In *Taxation in Developing Countries*, edited by Richard M. Bird & Oliver Oldman, 367–87. Baltimore, MD: Johns Hopkins University Press, 1990.

Cnossen, Sijbren. "Role and Rationale of Excise Taxation." In *Excise Tax Policy and Administration in Southern African Countries*, edited by Sijbren Cnossen, 1–20. Pretoria: University of South Africa, 2006a.

Cnossen, Sijbren. "The Role and Rationale of Excise Duties in ASEAN Countries." In *Excise Taxation in Asia*, edited by Stephen L.H. Phua, 1–24. Singapore: Faculty of Law, National University of Singapore, 2006b.

Cockfield, Arthur J. "Protecting the Social Value of Privacy in the Context of State Investigations Using New Technologies." *U.B.C. Law Review* 40 No. 1 (2007): 41–68.

Coleman, Cynthia. "Tax Refund versus Tax Return?" *Australian Tax Forum* 22 No. 2 (2007): 49–64.

"Customs and Biometrics: Security Systems of the 21st Century." *World Customs Organization News* (February, 2006): 40.

Das-Gupta, Arindam & Ira Gang. "Value Added Tax Evasion, Auditing and Transactions Matching." In *Institutional Elements of Tax Design*, edited by John McLaren, 25–48. Washington, D.C.: World Bank, 2003.

Das-Gupta, Arindam & Dilip Mookerjhee. *Incentives and Institutional Reform in Tax Enforcement*. New Delhi: Oxford University Press, 1998.

Das-Gupta, Arindam, D. Mookerjhee & D.P. Panta. *Income Tax Enforcement in India: A Preliminary Analysis*. New Delhi: National Institute of Public Finance and Policy, 1992.

De Soto, Hernando. *The Other Path*. New York: Perseus Books, 1989.

De Wulf, Luc & Jose B. Sokol, eds. *Customs Modernization Handbook*. Washington, D.C.: The World Bank, 2003.

Dhillon, Amardeep & Jan G. Bouwer. "Reform of Tax Administration in Developing Nations." In *Revenue Matters: A Guide to Achieving High Performance Under Taxing.*

Circumstances, edited by Brian Dustrud & Lisa Barr, 2005. Available at http://www.revenueproject.com.

Dolivio, F. "The IBM Secure Trade Lane Solution." *ERCIM (European Research Consortium for Information and Mathematics) News* 68 (January, 2007): 45.

Ebrill, Liam, Michael Keen, Jean-Paul Bodin & Victoria Summers. *The Modern VAT*. Washington, D.C.: International Monetary Fund, 2001.

Engelschalk, Michael, Samia Melhem & Dana Weist. *Computerizing Tax and Customs Administration*. PREM Note No. 44. Washington, D.C.: World Bank, 2000.

Engelschalk, Michael. "Creating a Favorable Tax Environment for Small Business." In *Taxing the Hard-to-Tax: Lessons from Theory and Practice*, edited by James Alm, James, Jorge Martinez-Vazquez & Sally Wallace, 275–312. Amsterdam: Elsevier, 2004.

Eschet, Gal. "FIPs and PETs for RFID: Protecting Privacy in the Web of Radio Frequency Identification." *Jurimetrics Journal* 45 No. 3 (Spring, 2005): 301–32.

Evans, Chris. "Studying the Studies: An Overview of Recent Research into Taxation Operating Costs." *eJournal of Tax Research* 1 No. 1 (2003): 64–92.

Foreign Investment Advisory Service (FIAS). *E-government in FIAS Projects— Case Study Compendium*. Washington, D.C.: World Bank, 2006.

Fjeldstad, Odd-Helge. "Fighting Fiscal Corruption: The Case of the Tanzania Revenue Authority." CMI Working Paper. Bergen: Christian Michelsen Institute, 2002.

Fox, William F. & Tami Gurley. "An Exploration of Tax Patterns Around the World." *Tax Notes International* (February 28, 2005): 793–807.

Gill, Jit B.S. *A Diagnostic Framework for Revenue Administration*. Washington, D.C.: World Bank, 2000.

Goolsbee, Austan. "The TurboTax Revolution: Can Technology Solve Tax Complexity?" In *The Crisis in Tax Administration*, edited by Henry J. Aaron & Joel Slemrod, 124–38. Washington, D.C.: Brookings Institution Press, 2004.

Gordon, Roger & Wei Li. "Tax Structure in Developing Countries: Many Puzzles and A Possible Explanation." *Journal of Public Economics*, 93 (August 2009), Nos. 7-8: 855-66.

Gray, John & Emma Chapman. *Evaluation of Revenue Projects Synthesis Report, Volume 1*. London: Department for International Development, 2001.

Harberger, Arnold C. "Lessons of Tax Reform from the Experiences of Uruguay, Indonesia, and Chile." In *Tax Reform in Developing Countries*, edited by Malcolm Gillis. Durham, NC: Duke University Press, 1989.

Hart, Oliver, Andrei Schleifer & Robert Vishny. "The Proper Scope of Government: Theory and Application to Prisons." *Quarterly Journal of Economics* (November, 1997): 1127-61.

Hettich, Walter & Stanley Winer. *Democratic Choice and Taxation*. New York: Cambridge University Press, 1999.

Inter-American Development Bank (IDB). "Recommendations and Best Practices on Taxation of SME in Latin America." Washington, D.C.: Inter-American Development Bank, 2007.

Information Technology for Adoption and Intelligent Design for e-Government (ITAIDE).*eTaxation & eCustoms Demonstrator D5.1.4*. Project No. 027829. April 17, 2007. Available at ttp://www.itaide.org/Projects/434/ITAIDE%20deliverables/Del%205%201%204%20BLL%20eTaxation%20and%20eCustoms%20demonstrator.pdf.

Jain, Sadhana. "Remote Sensing Application for Property Tax Evaluation." *International Journal of Applied Earth Observation and Geoinformation* 10 No. 1 (February, 2008): 109-21.

Jenkins, Glenn P., ed. *Information Technology and Innovation in Tax Administration*. The Hague: Kluwer Law International, 1996.

Jenkins, Glenn P., Chun-Yan Kuo & Keh-Nan Sun. *Taxation and Economic Development in Taiwan*. Cambridge, MA: John F. Kennedy School of Government, 2003.

Keen, Michael. "VAT, Tariffs and Withholding: Border Taxes and Informality in Developing Countries." IMF Working Paper No. 07/174. Washington, D.C.: International Monetary Fund, 2007.

Keen, Michael & Ben Lockwood. "Is the VAT a Money Machine?" *National Tax Journal* 49 No. 4 (December, 2006): 905-28.

Kelly, Roy. "The Evolution of a Property Tax Information Management System in Indonesia." In *Information Technology and Innovation in Tax Administration*, edited by Glenn Jenkins, 115-35. The Hague: Kluwer Law International, 1996.

Kenny, L.W. & S.L. Winer. "Tax Systems in the World: An Empirical Investigation into the Importance of Tax Bases, Administration Costs, Scale and Political Regime." *International Tax and Public Finance* 13 No. 2/3 (2006): 181-216.

Kirchler, Erich, Erik Hoelzl & Ingrid Wahl. "Enforced Versus Voluntary Tax Compliance: The 'Slippery Slope' Framework." *Journal of Economic Psychology* 29 No. 2 (April, 2008): 210-25.

Kumar, Sanjay, A.L. Nagar & Sayan Samanta. "Indexing the Effectiveness of Tax Administration." *Economic and Political Weekly* (December 15, 2007): 104-10.

Litvack, Jenny, Junaid Ahmad & Richard M. Bird. *Rethinking Decentralization.* Washington: World Bank, 1998.

Manasian, David. "Digital Dilemmas: A Survey of the Internet Society." *The Economist* 366 No. 8308/23 (Special Section, January, 2003): 25.

Manglik, Gauri. "Using Technology to Reduce Income Tax Evasion in India." *Tax Notes International* 50 No. 8 (May 26, 2008): 705-14.

Mann, Arthur. *Are Semiautonomous Revenue Authorities the Answer to Tax Administration Problems in Developing Countries? A Practical Guide.* Washington, D.C.: United States Agency International Development, 2004. Available at http://www.fi scalreform.net/best_practices/pdfs/sara_study_fi nal_jan-4-2005.pdf.

McLure, Charles & Santiago Pardo. "Improving the Administration of the Colombian Income Tax." In *Improving Tax Administration in Developing Countries,* edited by Richard M. Bird & Milka Casanegra, 124-44. Washington, D.C.: International Monetary Fund, 1992.

Minow, Martha. "Public and Private Partnerships: Accounting for the New Religion." *Harvard Law Review* 116 No. 5 (March, 2003): 1229-70.

Monarch Products and Services. *Tracking Reusable Metal Shipping Containers Using Passive RFID.* Avery Dennison White Paper. Miamisburg, OH: Avery Dennison, 2007. Available at http://www.integratedsolutionsmag.com/index.php?option=com_docman&task=doc_view&gid=86

Moore, Mick. "How Does Taxation Affect the Quality of Governance?" *Tax Notes International* 47 No. 1 (July 2, 2007): 79-98.

Moya, Rafael & José-Damián Santiago. "Information Technology: Strategy of the Spanish Tax Administration." In *Improving Tax Administration in Developing Countries,* edited by Richard M. Bird & Milka Casanegra de Jantscher, 211-35. Washington, DC: International Monetary Fund, 1992.

Mushkin, Selma J. "Designing a Credit Card Experiment." In *Public Prices for Public Products,* edited by Selma Mushkin, 407-8. Washington, D.C.: The Urban Institute, 1972.

Organization for Economic Co-operation and Development (OECD). *Survey of Trends in Taxpayer Service Delivery Using New Technologies.* Paris: OECD, 2005.

Organization for Economic Co-operation and Development (OECD). Using Third Party Information Reports to Assist Taxpayers Meet Their Return Filing Obligations—Country Experiences with the Use of Pre–Populated Personal Tax Returns. Paris: OECD, 2006.

Peterson, Stephen B. "Making IT Work: Implementing Effective Financial Information Systems in Bureaucracies of Developing Countries." In *Information Technology and Innovation in Tax Administration,* edited by Glenn Jenkins, 177-94. The Hague: Kluwer Law International, 1996.

Pinhanez, Monica. "Reinventing VAT Collection: Industry Vertical Assessment, Revenue Increase, and Public Sector Reliability." Ph.D. dissertation. MIT, Cambridge, MA, 2008.

Plumley, Alan H. & C. Eugene Steuerle. "Ultimate Objectives for the IRS: Balancing Revenue and Service." In *The Crisis in Tax Administration,* edited by Henry J.

Aaron & Joel Slemrod, 311–38. Washington, DC: Brookings Institution Press, 2004.

Radian, Alex. Resource Mobilization in Poor Countries: Implementing Tax Reform. New Brunswick, NJ: Transaction Books, 1980.

Rädler, Albert J. "The Future of Exchange of Information." In *L'evoluzione dell'ordinamento tributario italiano*, edited by Victor Uckmar. Milan: CEDAM, 2000.

Rajaraman, Indira. "Taxing Agriculture in a Developing Country: A Possible Approach." In *Taxing the Hard-to-Tax: Lessons from Theory and Practice*, edited by James Alm, James, Jorge Martinez-Vazquez & Sally Wallace, 245–68. Amsterdam: Elsevier, 2004.

Ramírez Acuña, Luis Fernando. "Privatization of Tax Administration." In *Improving Tax Administration in Developing Countries*, edited by Richard M. Bird & Milka Casanegra de Jantscher, Washington, DC: International Monetary Fund, 1992.

Roller, Lars-Hendrik & Leonard Waverman. "Telecommunications Infrastructure and Economic Development: A Simultaneous Approach." *American Economic Review* 91 No. 4 (2001): 909–23.

Rutkowski, Jan. "Taxation of Labor." In *Fiscal Policy and Economic Growth: Lessons for Eastern Europe and Central Asia*, edited by Cheryl Grey, Tracey Lane & Aristomes Varoudakis. Washington, D.C.: World Bank, 2007.

Saltzman, Michael I. *IRS Practice and Procedure*. 2d ed. Boston: Warren, Gorham & Lamont, 1991.

Schwartz, Paul. "The Future of Tax Privacy." *National Tax Journal* 61 No. 4, Part 2 (December, 2008): 879–96.

Shoup, Carl S. *Public Finance*. Chicago: Aldine, 1969.

Sia, S.K. & B.S. Neo. "Reengineering Effectiveness and the Redesign of Organizational Control: A Case in Study of the Inland Revenue Authority of Singapore." *Journal of Management and Information Services* 14 (1997): 69–82.

Silvani, Carlos & Alberto H.J. Radano. "Tax Administration Reform in Bolivia and Uruguay." In *Improving Tax Administration in Developing Countries*, edited by Richard M. Bird & Milka Casanegra, 19–59. Washington, D.C.: International Monetary Fund, 1992.

Slack, Enid, Almos Tassonyi & Richard M. Bird. "Reforming the Property Tax in Ontario: A Never-Ending Story." International Tax Program Paper No. 0706. Toronto: University of Toronto, Rotman School of Management, 2007.

Slemrod, Joel. "Optimal Taxation and Optimal Tax Systems." *Journal of Economic Perspectives* 4 No. 1 (Winter, 1990): 157–78.

Slemrod, Joel. "Taxation and Big Brother: Information, Personalisation and Privacy in 21st-Century Tax Policy." *Fiscal Studies* 27 No. 1 (March, 2006): 1–15.

Slemrod, Joel & Shlomo Yitzhaki. "The Optimal Size of a Tax Collection Agency." *Scandinavian Journal of Economics* 89 No. 2 (September, 1987): 183–92.

Soos, Piroska. "Self-employed Evasion and Tax Withholding: A Comparative Study and Analysis of the Issues." *UC Davis Law Review* 24 No. 1 (Fall, 1990): 107–94.

Stern, Richard E. & Paul A. Barbour. "Designing a Small Business Tax System that Enhances Growth: Lessons from Africa." Washington, D.C.: World Bank, Foreign Investment Advisory Service, 2005.

Swedberg, Claire. "Wireless System Aims to Fast-Track Heineken From Holland." *RFID Journal* (October 30, 2006).

Swire, Peter P. "Financial Privacy and the Theory of High-Tech Government Surveillance." *Washington University Law Quarterly* 77 No. 2 (Summer, 1999): 461–512.

Taliercio, Roberto. "Designing Performance: The Semi-Autonomous Revenue Authority Model in Africa and Latin America." World Bank Policy Research Working Paper No. 3423. Washington, D.C.: World Bank, 2004.

Toro, Juan. "Implementing VAT: Problems and Experiences." Presentation at the First Global ITD Conference on VAT. Rome, March 15–16, 2005. Available at http://www.itdweb.org/VATConference/documents/Presentations/Parallel3_Implementing%20VAT_Problems%20&%20Experiences%20_JToro.ppt.

Treblicock, Michael J. & Edward M. Iacobucci. "Privatization and Accountability." *Harvard Law Review* 116 No. 5 (March, 2003): 1422–53.

United Nations, Department of Economic and Social Affairs. *Compendium of Innovative E-government Practices*. New York, 2007.

Vazquez-Caro, Jaime & Joel Slemrod. "Issues in Colombian Tax Administration." In *Fiscal Reform in Colombia: Problems and Prospects*, edited by Richard M. Bird, James M. Poterba & Joel Slemrod, 139–52. Cambridge, MA: MIT Press, 2005.

Vehorn, Charles L. & John Bondolo. "Organizational Options for Tax Administration." *Bulletin for International Fiscal Documentation* 53 No. 11 (November, 1999): 499–512.

Werbach, Kevin. "Sensors and Sensibilities." *Cardozo Law Review* 28 No. 5 (April, 2007): 2321–71.

Whitman, James Q. "The Two Western Cultures of Privacy: Dignity versus Liberty." *Yale Law Journal* 113 No. 6 (April, 2004): 1151–221.

World Bank and International Finance Corporation. *Doing Business in 2008*. Washington, D.C.: World Bank, 2008.

World Bank and International Finance Corporation. *Designing a Tax System for Micro and Small Businesses: Guide for Practitioners*. Washington, D.C.: World Bank, 2007.

CHAPTER 5
Information Technology in the U.S. Tax Administration

Andrew T. Buckler

§5.01 INTRODUCTION

Tax administration is primarily an information management business. At the U.S. Internal Revenue Service (IRS) and other tax agencies, massive amounts of data are received and processed every year in the form of tax returns, payments, information documents and many other types of submissions. That information is supplied by many entities, including individual and business taxpayers, tax professionals, financial institutions, payroll servicing firms and other government agencies at the federal, state and local levels. The information environment is high volume and extremely complex from both a business and technological perspective. For example, IRS workload in Fiscal Year (FY) 2010 included:[1]

- 230 million tax returns of all types, from single page 1040 EZ's to corporate returns that can be thousands of pages;
- USD 2.3 trillion in total collections;
- 122 million refunds totaling USD 467 billion;
- 2.7 billion third-party information documents;
- 72 million customer service phone calls;
- 304 million visits to the IRS.GOV website;
- 8.4 million math error cases;
- 1.7 million returns examined.

The massive amounts of information generated by these activities are processed by the core systems of the IRS to update and maintain account status, provide service to

---

1. Internal Revenue Service Data Book, 2010; Publication 55B; Washington, DC; March 2011.

citizens and drive compliance programs. The technology environment is also highly complex with hundreds of applications running on a wide variety of hardware platforms used by over 94,000 IRS employees, as well as by the taxpaying public.

Tax administration agencies have relied heavily on information technology (IT) throughout the years to improve the efficiency and effectiveness of their operations. IRS was an early governmental pioneer in IT and has often been at the forefront of federal government technology implementations. Throughout the past few decades, the focus of agency IT has been to reduce costs and improve the effectiveness of their business operations.

With the explosion of IT capabilities in all aspects of modern life, tax agencies face many new challenges. Significant portions of the world's economic activity are now conducted via the internet. Instantaneous communications and broader participation in complex financial transactions have combined to enable a rapid globalization of activity, bringing new players into the international financial community. The increasing speed, complexity and global scope of economic transactions challenge the ability of tax agencies to understand and monitor activity for tax implications.

Over the past few years, we have seen the impact of some players in this community who are actively promoting tax evasion schemes, such as hiding assets through the use of off-shore financial institutions and structures. Tax agencies are challenged to obtain and analyze information to help them to identify and address such schemes.

In addition, the expectations of the public have changed with regard to the speed and manner in which services are provided. As private sector companies have upgraded their customer service capabilities and increasingly moved their operations to the web, government agencies are challenged to provide similar types of services.

In the midst of all this change, the IRS is also faced with a number of other challenges— existing tax law requires significant annual changes in processing systems; each year, new legislation and regulations continue to add complexity to the tax code; new responsibilities (such as, economic stimulus payments and health care reform) have been assigned to the agency; cybersecurity threats are constantly evolving; and large federal budget deficits have increased pressures to reduce operating costs.

To understand how the IRS and other tax agencies will respond to these opportunities and challenges, it is helpful to start with a look at the current IT environment in the context of the long-term evolution life cycle.

§5.02 THE EVOLUTION OF INFORMATION TECHNOLOGY IN TAX ADMINISTRATION

As stated above, tax agencies in general, and IRS specifically, were early government technology adopters, and have continually expanded their IT capabilities over the years as computing power and capabilities have matured. Many of their early systems remain in production, while new ones have been implemented over the years. As a result, the IT environment includes systems that were constructed at different times, using different technology platforms, different programming languages and providing different types of functionality. This leads to a very complex operational environment, and sometimes limits the capabilities that can be added to existing systems.

This situation is by no means unique to the IRS or other tax agencies. Many government and private sector organizations that have been in place for many years are in very similar positions. Understanding these systems and their role in tax administration can provide some insight to where agencies must go in the future.

[A] Early Systems Focused on Processing Incoming Data and Maintaining Accounts

One of the first major IRS IT deployments was the Individual Master File (IMF) in 1963. The IMF is still in production in 2011 (although the Customer Account Data Engine 2 program is underway to replace it) and serves as the core authoritative system of record for individual taxpayer accounts. It was followed over the years by other master files for businesses, tax exempt organizations, employee plans, etc.

The IMF utilized the technology of its era— flat, sequential data files on magnetic tape, processed through assembler language code. Automation of the master files eliminated the need for paper records and file folders that were stored in local offices all over the country. The master files also automated a number of work processes, including certain calculations and the generation of notices, reducing the need for employees to perform these tasks by hand. By eliminating the need for paper account files, IMF enabled IRS to gain efficiencies of scale by centralizing its processing operations from dozens of local offices into ten service centers in the 1960s.

By the early 1970s, the IRS implemented new capabilities to allow employees to directly access and modify taxpayer accounts. These new systems took advantage of technology improvements at the time, including the use of database technology and COBOL coding. The system was also built to monitor activity on taxpayer accounts and even take actions (i.e., issue notices, assess interest, reassign cases, etc.) depending on activity or lack of activity on the account. The Integrated Data Retrieval System (IDRS) is still the workhorse of the IRS, serving as the core case management system and providing the primary interface for employees to access taxpayer accounts.

[B] Efficiency Is Driven by the Automation of Work Processes

Having established the tax administration technology foundation around account maintenance, the next wave of automation focused on providing automated tools to IRS employees in the customer service and compliance functions. Again, these systems were implemented over many years, and each one was designed around the technology capabilities of its time.

Initially, the focus was on automating tasks that were performed by employees, such as performing calculations, generating reports, issuing letters, capturing management information, etc. As technological capabilities evolved, these end-user systems changed as well, providing capabilities such as automated case selection, assignment and routing, systemic account monitoring, system-generated actions and document imaging to enable fully electronic case files.

A very successful example of this phase of automation was the Automated Collection System (ACS), deployed in the mid-1980s. ACS dramatically changed the IRS Collection function. Automated case files eliminated the old business process that was dependent on the maintenance of millions of case folders maintained in local offices, eventually enabling employees anywhere in the country to work any case. ACS also integrated information from multiple IRS databases to automate the generation of enforcement actions like the filing of liens and levies.

Business intelligence was introduced to the system, implementing decision rules developed through years of experience to prioritize cases, so that as case workers became available, they would automatically receive the "next best case" on their workstations. ACS was also integrated with the toll-free telephone system, so that when a taxpayer called, their case information would automatically be provided to the case worker taking the call. Predictive dialer technology was integrated with ACS to automate the generation of outgoing phone calls, again using business intelligence decision rules to determine the next best case.

ACS ranks as a major success story for the use of automation in tax administration. IRS has continued to enhance the system, and it remains a mainstay of the IRS Collection process. Systems with similar work management and processing capabilities were deployed for other IRS customer service and enforcement functions.

[C] Automation and Electronic Filing Drive Significant Processing Cost Reduction

Since information is the primary business of tax administration, agencies are faced with the task of receiving massive amounts of data and getting that data into their systems. For many years, there was no choice but to receive that data on paper and use employees to transcribe the information to machine readable formats. The IRS operated ten service center "pipelines" with thousands of employees dedicated to data transcription and perfection. This was a very costly operation, prone to frequent mistakes, both by taxpayers and by IRS employees.

IRS and other tax agencies embarked on a number of approaches to reduce these costs and streamline the process. Optical character recognition technology was used for processing of some paper documents, but eventually, the IRS focused on enabling the direct electronic receipt of data as much as possible. Systems for electronic submission of third-party information documents (W-2's, 1099's, etc.) have been very successful, with over 88% submitted electronically in FY 2010.[2] Electronic payment processing was enabled through the Electronic Funds Transfer Payment System, which processes most federal tax deposits of employee withholding and payroll taxes, along with many other electronic payments.

---

2. Ibid.

However, the most important IRS program in this arena was the implementation of electronic filing of tax returns. The IRS strategy for electronic filing was to partner with stakeholders in the tax preparation industry (both practitioners and software providers) to drive growth in electronic filing. Today, IRS offers electronic filing for a wide variety of forms and schedules for individual taxpayers, businesses and exempt organizations. In FY 2010, 116 million returns of all types were filed electronically, including over 98 million individual returns (more than two-thirds of the total),[3] a remarkable level of market penetration for a government program. The success of the electronic filing program has reduced costs, while improving the accuracy of tax return processing.

[D] Information Matching Programs Provide Significant Compliance Presence

The IRS Information Return Program (IRP) was implemented in the early 1970s. Under this program, payors (employers, financial institutions, etc.) are required to provide the IRS with reports (W-2's, 1099's) on many types of payments that they make to individual taxpayers. During FY 2010, IRS received more than 2.7 billion IRP documents.[4]

IRS matches those reports against the income reported on tax returns and generates compliance cases where mis-matches are detected. Information is also shared with many state tax agencies. This program has been hugely successful, both in terms of the direct revenue collected from the identified mis-match cases, but also in serving as a deterrent. Information return matching programs generated over 4.3 million automated under-reporter cases, with a total of USD 7.2 billion in additional assessments in FY 2010,[5] providing a broad compliance presence and significant revenue generation at relatively low cost.

The success of these programs has led policy makers to seek ways to expand the types of income that are subject to reporting, as well as requiring additional information to help the IRS to systemically identify noncompliance. Recent legislation has added information reporting requirements for the basis of stock transactions, certain payments between businesses, merchant credit card payments and reporting of foreign financial accounts.

[E] Web-Based Systems Offer Improved Service to Taxpayers

In recent years, tax agencies have begun to offer citizens new capabilities enabled by internet-based technology. Most agencies have public facing websites, similar to IRS.GOV, which serves as a major source of information for the public and for tax

---

3. *Ibid.*
4. *Ibid.*
5. *Ibid.*

professionals to understand and navigate the U.S. tax system. IRS.GOV received over 1.6 billion page views in FY 2010.[6]

IRS provides the "Where's My Refund?" self-service web application which was used 66 million times by taxpayers to track the status of their tax refund during FY 2010.[7] The underlying technology was re-used to deploy "Where's My Economic Stimulus Payment?" which processed 58 million taxpayer requests during FY 2009.[8] The use of this technology to provide self-service assistance to the public was critical to the Service's successful implementation of the Economic Stimulus Payment program, since the customer service phone lines could easily have been overwhelmed without it.

"Where's My Refund?" has been so successful that for filing season 2011, IRS released it as the agency's first mobile application for the iPhone and Android mobile phones. In addition to providing better service to taxpayers, self-service applications help the agency to manage its customer service workload by drawing a large number of potential contacts away from the phones, allowing IRS to focus its customer service representatives on more complex taxpayer inquiries that require direct interaction.

Many states have also constructed internet applications to provide services to their taxpayers, and some have gone farther than the IRS by offering free electronic return filing directly on their websites. IRS has chosen not to offer this service directly, but rather to continue to partner with the tax preparation industry to support electronic filing. However, IRS has worked with the industry to establish the Free File Alliance. Members of the Alliance offer free tax preparation software and electronic filing to taxpayers with relatively simple returns and incomes below certain thresholds. Almost three million taxpayers took advantage of Free File in FY 2010.[9]

IRS has begun to offer additional services to taxpayers via the web. For example, certain taxpayers who have an unpaid tax liability to the IRS can take advantage of the On-line Payment Agreement (OPA) application, which enabled over 61,000 taxpayers in FY 2010[10] to enter into an installment agreement to pay their tax debt without ever speaking to an IRS employee. Applications such as OPA— which allow access to greater amounts of account information and allow taxpayers to make transactions online— require a higher level of data security and identity-proofing. This is a major issue for the future of expanded online services as discussed below.

§5.03 PLANNING FOR THE FUTURE: DEVELOPING AN IT PROGRAM ROADMAP

Tax agency Chief Information Officers (CIOs) face a daunting array of challenges for the future, many of which are discussed later. Top priority must always be given to the continued successful operation of the installed base of systems to achieve current business objectives and to maintain public confidence in the effectiveness of the tax

---

6. Ibid.
7. Ibid.
8. Internal Revenue Service Data Book, 2009; Publication 55B; Washington, DC; March 2010.
9. Internal Revenue Service Data Book, 2010; Publication 55B; Washington, DC; March 2011.
10. Ibid.

administration system. However, there has been and will continue to be heavy demand to deliver new capabilities for both internal and external customers, even as agency IT budgets are expected to be relatively flat or even shrinking over the next several years.

The IRS and other tax agencies are resource limited, both from a funding perspective and in terms of their internal capacity to manage change. As a result, tough decisions must be made to prioritize the allocation of IT resources to assure that the key strategic objectives of the agency are being addressed. Effective IT planning cannot be the sole responsibility of the CIO. IT investment must be driven by the strategic business plans of the agency. This requires effective partnerships among agency business leadership, the CIO and external stakeholders (including the executive branch, legislative oversight, tax industry representatives and taxpayer groups).

Processes must be implemented to make sound IT investment decisions, set strategic and tactical priorities and outline investment plans via a well-defined IT program roadmap. To accomplish this effectively, tax agency CIOs must:

- partner with business leadership and other stakeholders to develop a clear, realistic strategic technology vision for tax administration. This is an iterative process where business priorities, goals and objectives must be balanced off against technical capabilities and resource availability (both funding and people);
- establish disciplines within their own organizations to effectively define solutions, costs and timelines, so that strategic investment decisions can be based on reliable information;
- identify foundational investments necessary to enable other capabilities; for instance, systems and processes for electronic identity management and authentication, electronic signatures and secure electronic messaging are needed to enable a wide variety of expanded web-based services for the public;
- establish a plan for maintaining and refreshing the technology infrastructure that supports current operations;
- define a series of time-phased architectural views from both business and technology perspectives;
- develop a realistic program roadmap to sequence projects and programs over time while recognizing resource limitations (both funding and people/skills);
- gain broad stakeholder (both internal and external) understanding of and commitment to the roadmap;
- implement disciplined project/program management and governance processes to oversee successful delivery of planned projects and programs.

IRS has taken major steps forward in recent years to implement these processes. In 2007, the Service published its Modernization Vision & Strategy,[11] which described its

---

11. Internal Revenue Service IT Modernization Vision & Strategy, 2nd Edition, Document 12417, Washington, DC, October 2007.

investment decision-making processes, and provided a roadmap for future information technology investments.

§5.04 TECHNOLOGY CHALLENGES FACING TAX AGENCIES FOR THE FUTURE

Information technology continues to evolve, posing both challenges and opportunities for the IRS and other tax administration agencies. Expanded computing power, improved data storage technologies, new software capabilities, high speed data communications, mobile computing and social media all open up exciting new possibilities to improve operations, reduce costs and bring expanded business capabilities to both internal users and the public. With those possibilities comes the burden of high expectations for tax agencies to provide services comparable to private sector firms.

The other challenge posed by the rapid progression of technology is the impact on financial activity and transactions. Technology has enabled whole new areas of commerce, changed the way that financial transactions are initiated and executed, and created new methods of payment. The internet has enabled an explosion of financial activity that crosses international boundaries, posing a huge challenge for tax agencies to identify, understand and track these transactions to assure compliance with the tax laws.

Addressing these challenges will require significant investment to upgrade agency systems and software. As noted earlier, tough decisions will have to be made as to where limited resources will be invested to drive value for the IRS and its stakeholders. Major investment decisions can be grouped into six broad categories.

[A] Support Delivery of Current Operations

For many citizens, their only direct interaction with the federal government over the course of a year is the filing of their income tax return. Many households depend on receiving their refund checks each year to make ends meet. IRS issued over 119 million refunds to individual taxpayers, totaling more than USD 358 billion in FY 2010.[12] The issuance of tax refunds is an important factor in the economic activity of the nation. As a result, public perception of the IRS and to some extent, the government as a whole, is dependent on the IRS' ability to efficiently and timely process tax returns and generate refunds.

For many years, the top priority of the IRS entering each fiscal year has been the successful delivery of operations during the "filing season." That focus has paid off—IRS has delivered successful filing seasons for the past 25 years, despite the wide variety of obstacles and challenges it has faced. Over the past few years, these challenges have included significant tax legislation that was passed into law as late as mid-December, along with special requirements beyond the Service's traditional

---

12. Internal Revenue Service Data Book, 2010; Publication 55B; Washington, DC; March 2011.

mission, such as the issuance of economic stimulus payments. These and other challenges have required very significant and rapid IRS responses from both an operational and technology perspective.

In addition to filing season return processing, it is critical that the Service assure delivery of its many other mission activities. Effective customer service and compliance programs are essential to maintaining taxpayer confidence that the tax system is administered fairly and efficiently. Loss of that confidence could seriously undermine the voluntary compliance basis of the US tax system.

As the IRS and other tax agencies develop their IT investment plans, first consideration must be given to assuring the delivery of current operations. Thus, resources must be devoted to operation and maintenance of production systems. In addition, plans must include sufficient flexibility and resource availability to respond to late legislation and new requirements that may be thrust upon the agency.

Another critical requirement that must be factored into the planning process is the need to maintain, operate and refresh the agency's massive technology infrastructure. IRS operates three major data centers, including multiple mainframe computer platforms and thousands of servers. The Service's nationwide data network supports access to systems and data from the workstations of over 94,000 employees in hundreds of locations across the country. The Service's toll-free telephone network infrastructure processed nearly 72 million phone calls in FY 2010, with approximately 35 million handled through automated systems, and another 37 million routed to live assistants in two dozen call centers around the country.[13] IRS also has an extensive presence on the web, with the IRS.GOV website providing over 220 million downloads during FY 2010.[14]

The IRS has established a planning process to provide for regular refreshment of its IT infrastructure, a critical requirement to assure the continuity of current operations. As budget pressures force agencies to seek cost reductions, it is critical that capital infrastructure investments are maintained to assure the viability of IT systems in the current production environment.

[B] Carry Out New Responsibilities

In recent years, Congress has used the tax administration system and the operational capabilities of the IRS to implement a number of programs that have extended the traditional core mission of the agency. The Earned Income Tax Credit (EITC) is essentially an income redistribution program, accomplished through the tax code. The First Time Homebuyer's Tax Credit (FTHTC) was used as a means to stimulate the economy during the recession. During FY 2010, the FHTHC was claimed on 2.2 million tax returns, for credits totaling over USD 15 billion.[15]

Effective administration of provisions such as EITC and FTHTC requires IRS to perform outreach and educational activities to assure that eligible citizens, who might

---

13. *Ibid.*
14. *Ibid.*
15. *Ibid.*

not otherwise be required to file tax returns, are informed of the credits and how to claim them. Refundable credits such as these also pose compliance challenges to the IRS, since they are prime targets for refund fraud. As a result, implementation of these provisions required special programming to modify multiple IRS systems to reach out to potentially eligible taxpayers, process the new credit, screen refunds for potential errors and fraud, and enable post-filing compliance systems to address the new requirements.

IRS has been asked to take on new responsibilities that go beyond its traditional tax administration mission. For example, in FY 2009 the Service issued over 26 million economic stimulus payments.[16] The Patient Protection and Affordable Care Act (PPACA) health care reform legislation places many new demands on the Service. Over the next few years, IRS will have to establish processes and systems to exchange new types of information with other federal agencies, state governments, insurance exchanges and other participants in the health care industry. New submission processing, customer service and compliance programs must be planned and implemented. These new business operations will have significant impact on existing IRS systems, and will generate the need for new technology capabilities.

Implementation of these new programs will be a significant challenge from both an operational and technology perspective, and will require a very significant investment of both financial and human resources. As the agency plans its technology future, the Service must do its best to provide the flexibility and agility to respond to the new responsibilities that Congress may assign in coming years.

[C]   Protect Sensitive Data

Protection of the security and privacy of tax data has been a high priority for the IRS for many years. The agency has long recognized that maintaining public confidence that sensitive personal information is secure and protected is critical to the US voluntary compliance system. Tax data is subject to very strict disclosure rules, and IRS has worked hard to assure the protection of its data.

However, the challenges associated with systems security and data protection continue to grow for tax agencies, just as they do for other government and private sector organizations. There are two major reasons: First, cyber threats are becoming increasingly organized, sophisticated and dangerous. Second, as more and more transactions and information are made available via web-based applications, new opportunities for malicious activity are created.

Enabling the expansion of the volume and types of transactions conducted with taxpayers via the internet requires the implementation of a number of core foundational capabilities. Strong identity and access management systems and practices must be implemented, so the IRS can be assured that the person conducting a transaction is who they say they are. Identity management for over 140 million individual and 10

---

16.   Internal Revenue Service Data Book, 2009; Publication 55B; Washington, DC; March 2010.

million business filers[17] presents a massive challenge, especially since most taxpayers do not have a need to conduct electronic transactions with IRS on a regular basis. Other capabilities that are needed to enable expanded web-based capabilities include electronic signatures and secure methods for taxpayers to send and receive electronic messages and documents with the IRS.

Tax agencies will need to continue to upgrade their cybersecurity technology and practices, and assure that all new applications and business capabilities include appropriate protections and controls. These requirements will continue to require significant investment of resources in coming years.

[D] Drive Business Efficiency and Cost Reduction

Government agencies at all levels are coming under increasing pressure to control and reduce costs. Even though tax agencies deliver the revenue that keeps the rest of government running, there will be an expectation to take every step possible to control costs while improving business results. From a technology perspective, cost control has two aspects: controlling the cost of IT operations; and providing new capabilities to enable agency business functions to improve their productivity.

Like their counterparts in all government organizations, tax agency CIOs will need to implement new technologies and practices to reduce the cost of operating the installed base of systems and applications. The federal government has instituted a number of government-wide initiatives to address IT cost reduction, including data center consolidation, the use of cloud computing and software as a service (SaaS) technologies. The IRS and other tax agencies will continue to explore the opportunities presented by these and other new technologies to reduce costs while maintaining or improving services and security.

In the systems operations arena, IRS and many other tax agencies spend a significant amount of resources to operate and maintain the many applications and systems in their environments. Consolidation of multiple stove-piped applications, movement to more modern standardized hardware and software platforms, and retirement of old applications all provide potential opportunities to reduce operational costs through standardization and simplification of the IT environment. However, there are initial investment costs necessary to achieve the potential long-term savings, and the benefits of these investments will have to be balanced against other investment opportunities that would deliver new business capabilities.

The second area for IT investment to improve cost efficiency is in supporting tax agency business functions by providing new and improved capabilities. Staffing levels at the IRS declined by approximately 2% between FY 2000 and FY 2010, while total returns filed have increased by 2%.[18] As a result, IRS business operations are constantly under pressure to find ways to do more with less, and technology is a prime opportunity for achieving that goal.

---

17. Internal Revenue Service Data Book, 2010; Publication 55B; Washington, DC; March 2011.
18. Internal Revenue Service Data Book, 2000; Publication 55B; Washington, DC; 2001.
    Internal Revenue Service Data Book, 2010; Publication 55B; Washington, DC; March 2011.

IRS has already achieved a number of very successful technology implementations that have driven significant business productivity improvements, some of which were described earlier. One of the most visible has been the successful expansion of electronic return filing. Having achieved a total of over 116 million returns of all types filed electronically in FY 2010,[19] IRS has been able to significantly reduce its workforce dedicated to input processing. As a result, the Service has eliminated returns processing "pipeline" operations at six of the ten processing centers where that work was previously performed.

While almost all business functions of the IRS have been improved by IT systems over the years, opportunities remain to add capabilities to existing applications or to invest in new systems to improve business productivity and effectiveness. Budget pressures, along with growing workload demands, will continue to drive investments to improve operational efficiency and reduce costs.

[E] Improve Service to Taxpayers and Tax Professionals

In 2006 and 2007, the IRS published two volumes of the Taxpayer Assistance Blueprint (TAB).[20] These documents reported the results of research into the needs and preferences of taxpayers in their interactions with IRS, and outlined strategies and actions to improve both the effectiveness and efficiency of services.

The research found that approximately 41% of all taxpayers reported that they contacted IRS for assistance over a one- to two-year period, and taxpayers seeking assistance contacted the Service an average of over 4 times. The report also states that taxpayers' preference for different communications channels (in person, telephone, web) changes, based on the type of interaction and the quality of the service provided.[21]

One very important opportunity for both service improvement and cost reduction is through the implementation of online self-help service to taxpayers. The first objective for service improvement in the TAB report states that the IRS should "maximize the taxpayer and partner value of the IRS website, making the electronic channel the first choice of taxpayers and partners for obtaining the information and services they need to comply with their tax obligations."[22]

IRS provides a great deal of content via their IRS.GOV website, including access to IRS forma and publications, along with a wide variety of information targeted to specific taxpayer populations, including individuals, businesses, tax exempt organizations and government entities. News and information is also provided for the tax professional community. While IRS.GOV has been very successful and won a number of accolades in prior years, there is a continuing need to update the technology and

---

19. Internal Revenue Service Data Book, 2010; Publication 55B; Washington, DC; March 2011.
20. The 2006 Taxpayer Assistance Blueprint Phase 1; Publication 4525; Washington, DC; April 2006; The 2007 Taxpayer Assistance Blueprint Phase 2; Publication 4576; Washington, DC; April 2007.
21. *Ibid.*
22. *Ibid.*

content delivery to keep up with current standards. Examples include improving search capabilities and content management processes.

Along with the information content, the IRS currently offers a relatively limited suite of web-based tax account-related transactions. The "Where's My Refund?" and On-line Payment Agreement applications were cited earlier as examples. However, other capabilities to allow taxpayers to view their account status, request account transcripts and conduct other interactions via the web will require IRS to implement new levels of identity management and authentication.

Since the refund application discloses very little information (the approximate date when a refund can be expected), authentication can be enabled by requiring the taxpayer to provide "shared secrets": Social Security number, filing status and refund amount. However, applications that would disclose more sensitive information or allow a taxpayer to make some type of change to their account would require additional protection to provide adequate certainty that the individual is who they say they are.

In situations where the IRS is initiating the interaction— e.g., with the On-line Payment Agreement application, where the web-based service is offered to the taxpayer through IRS correspondence— the additional factor can take the form of a Personal Identification Number (PIN) provided by the IRS. However, for transactions that would be initiated by the taxpayer, such as a transcript request, adding a second factor to the authentication process is much more problematic. The IRS is currently working to solve this problem and implement a strong electronic identity management and authentication system. This capability is foundational to enable a much broader range of web-based services to both improve customer service and drive new efficiencies.

Another opportunity for technology-enabled taxpayer service improvement is through the implementation of integrated multi-channel services. Major private sector organizations have successfully employed these capabilities to drive efficiencies and to improve the overall user experience. An example is the use of "web chat" to allow a customer service representative to provide assistance to a taxpayer who is attempting to use a web-based service. Technologies like this could allow the IRS to make more effective and efficient use of its customer service workforce by dynamically allocating the total workload from both telephone and internet channels, while providing better service to the taxpaying public.

[F]   Address Noncompliance with the Tax Laws

Based on research conducted on Tax Year 2001 tax filings, the IRS estimated that the overall voluntary compliance rate was approximately 84%, and that the gross "tax gap"— the difference between what taxpayers should have paid and what they actually paid on a timely basis— was approximately USD 345 billion. After subtracting revenue from enforcement actions and other late payments, the net tax gap was estimated at USD 290 billion.[23]

The Government Accountability Office lists "Enforcement of Tax Laws" as one of its key high risk areas in the federal government.[24] IRS and the Treasury Department have come under significant pressure from both the executive and legislative branches in recent years to address the problem. Treasury has listed continued improvement in information technology as one of seven key components of its strategy to reduce the tax gap.[25]

Addressing noncompliance is a very complex undertaking, since there are so many dimensions to the problem. IRS defines three main components of the tax gap: Nonfiling (7.8%), underreporting (82.6%) and underpayment (9.6%).[26] Each of these areas can be further decomposed based on the type of tax (income, employment, estate, excise) and the type of taxpayer entity (individuals, corporations, partnerships, tax exempt organizations, government entities, estates, trusts). To address noncompliance within each type of tax and population segment, IRS must follow a series of fundamental steps, each of which is supported by technology:

> Identify potential noncompliance— which taxpayers may be involved and how are they noncompliant? This includes identifying taxpayers who are involved in somewhat "traditional" tax avoidance behaviors— simply omitting or understating income and/or overstating deductions— as well as those who take overly aggressive positions on interpretation of complex tax provisions and those are engaged in more complicated and sophisticated schemes to evade taxes.
>
> Select cases to work— the Service has limited compliance resources, so once a population of potentially noncompliant taxpayers has been identified, IRS must decide which cases will be selected for enforcement action. Decisions are made based on the likelihood and magnitude of potential issues, but consideration must also be given to maintaining a compliance presence across all population segments to provide a deterrent effect.
>
> Conduct enforcement actions— provide information, systems, tools and work processes to enable IRS enforcement case workers to conduct examination and collection activities at the lowest cost to deliver maximum results.

With regard to the third point, as discussed earlier, the IRS has deployed many applications and systems improvements over the years to help improve the efficiency and effectiveness of their compliance operations. The Automated Collection System was cited as a very successful example of these systems, but other IRS enforcement functions have also deployed systems to automate work processes to maximize the efficiency and effectiveness of their front-line caseworkers. These systems have resulted in significant productivity gains over the years. Additional opportunities certainly still exist to further improve the tools and work processes for compliance

---

23. Update on Reducing the Federal Tax Gap and Improving Voluntary Compliance; U.S. Department of the Treasury; Washington, DC; July 2009.
24. High Risk Series: An Update; U.S. Government Accountability Office; Publication GAO-09-271; Washington, DC; January 2009.
25. Update on Reducing the Federal Tax Gap and Improving Voluntary Compliance; U.S. Department of the Treasury; Washington, DC; July 2009.
26. Ibid.

employees, but potentially more significant opportunities remain in the areas of case identification and selection.

The task of identifying potentially noncompliant taxpayers is very complex, since the range of taxpayer types is so broad and the nature of noncompliance is so varied. The problem of identifying an individual or small business that is not filing required returns or simply omitting cash basis income is much different than identifying a high wealth individual or business using a complex array of pass-through entities and possible off-shore accounts to hide income from the IRS. And the issues associated with large multi-national corporations taking aggressive positions on very complex sections of the tax code presents even a different challenge.

To address this variety and complexity, IRS has developed a number of systems and processes to identify different types of potential noncompliance. The Electronic Fraud Detection System screens tax returns for potentially erroneous or fraudulent refund claims before the refund is issued. As discussed earlier, the Information Reporting Program matches third-party information reports (W-2's, 1099's, etc.) against tax return information to identify possible nonfiler and under-reporter cases. The Service has a system known as the Discriminate Function (DIF) which utilizes data from IRS research programs to score returns for the likelihood of under-reporting. And there are a number of other internal systems used by the IRS operating divisions to perform queries and searches for returns from specific taxpayer populations that meet various criteria that might suggest potential compliance issues.

However, there are limitations to the existing data, processes and systems that leave some significant opportunities for improvement in compliance case identification. One approach is to enhance the data available to the IRS by expanding the types of payments that are subject to information reporting requirements. IRS research has shown that "reporting compliance is highest where parties other than the taxpayer are required to file information reports and withhold taxes from payments made."[27]

Legislation in recent years added reporting requirements for credit card payments to merchants, for stock cost basis (both effective in Tax Year 2011), and for certain payments between businesses (starting in Tax Year 2012). In addition, the Foreign Account Tax Compliance Act (FATCA) requires foreign financial institutions doing business in the US to report certain information on assets held by US taxpayers, starting in 2013.

Expanding the universe of financial information that is reported to IRS, making more data available for matching programs and analysis, is a significant step forward in the effort to uncover and deter noncompliance. The new information reports listed above pose additional challenges for the IRS from a technology perspective, since matching for businesses will be a much more complex undertaking. Systems must be developed to correlate the new data against tax returns, adjust for fiscal year versus calendar year filers, address the complexities of related business entities and pass-throughs, and implement new logic to screen out false mis-matches and identify the most productive cases to be worked.

---

27. *Ibid.*

While expanded information reporting is an important tool, the IRS must find other ways to address more complex compliance issues where third-party information reporting is not available. Over the years, various schemes have come to light through which high asset individuals and businesses have attempted to evade tax obligations. As IRS has uncovered these arrangements, the agency has aggressively pursued enforcement actions. The most recent example is the use of off-shore banks and financial institutions to hide assets and income. FATCA institutes new reporting requirements with regard to foreign assets, but it certainly will not completely address the problem.

To meet these and other compliance challenges that are sure to develop in coming years, the IRS must look to other ways to recognize trends in financial activity that may suggest compliance issues. The IRS and other tax agencies will need to explore unconventional data sources beyond those that are reported directly by taxpayers and payors. There are parallels to the intelligence community which has learned in recent years how to access external information sources and use emerging technologies to sift through large amounts of data to make the connections that reveal patterns of behavior.

Once potentially noncompliant taxpayers have been identified, the IRS must select which cases it will actively work. Again, this is a difficult challenge. Selection processes could clearly be improved as evidenced by the significant percentages of IRS audits that resulted in no change in Fiscal Year 2010:[28]

- Correspondence examinations of individual taxpayers— 14%
- Field examinations of small corporations (non-1120S) with balance sheet assets up to USD 10 million— 30%
- Field examinations of large corporations (non-1120S) with balance sheet assets over USD 10 million— 27%

No change examinations are a significant productivity drain for the IRS, as well as an unnecessary cost and waste of time for compliant taxpayers. It would be unrealistic to expect that no change rates could ever approach zero, and IRS is constantly working to refine its case selection criteria, but this is clearly an area that offers an opportunity for major productivity and effectiveness improvements, along with reduction in taxpayer burden.

Recent advances in data management technologies provide new capabilities and tools to consolidate and analyze large volumes of data from different sources and in different formats. New analytical tools and processes, many driven by the needs of the intelligence community, make it possible to mine large amounts of unstructured data from multiple sources to identify patterns and provide insights that would previously have gone undetected. By taking advantage of new and different external information resources, and applying emerging technologies to improve its analytical capabilities, the IRS could improve its ability to identify noncompliant behavior earlier, and to

---

28. Internal Revenue Service Data Book, 2010; Publication 55B; Washington, DC; March 2011.

optimize the allocation of its scarce enforcement resources to achieve the highest degree of efficiency and effectiveness.

§5.05 CONCLUSION

For the Internal Revenue Service, and for tax agencies at every level, the rapid evolution of information technology will continue to be a double-edged sword, posing both major opportunities and significant challenges. Opportunities include potential improvements to the efficiency and effectiveness of agency business operations, the capability to offer new and expanded services to taxpayers, and the use of new data sources and analytical techniques to better identify and address compliance issues.

Challenges will be posed by the greater expectations of stakeholders to keep pace with the private sector in utilizing new IT capabilities and by the rapid changes in global financial activity enabled by technology. In addition, agencies will be asked to develop and deploy new technology capabilities with flat or even shrinking budgets. The IRS and other agencies will continue to face an array of tough investment decisions to ensure that they can continue to deliver current operations while investing for the future.

§5.06 REFERENCES

High Risk Series: An Update; U.S. Government Accountability Office; Publication GAO-09-271; Washington, DC; January 2009

Internal Revenue Service Data Book, 2000; Publication 55B; Washington, DC; 2001

Internal Revenue Service Data Book, 2009; Publication 55B; Washington, DC; March 2010.

Internal Revenue Service Data Book, 2010; Publication 55B; Washington, DC; March 2011.

Internal Revenue Service IT Modernization Vision & Strategy, 2nd Edition, Document 12417, Washington, DC, October 2007.

The 2006 Taxpayer Assistance Blueprint Phase 1; Publication 4525; Washington, DC; April 2006.

The 2007 Taxpayer Assistance Blueprint Phase 2; Publication 4576; Washington, DC; April 2007

Update on Reducing the Federal Tax Gap and Improving Voluntary Compliance; U.S. Department of the Treasury; Washington, DC; July 2009

PART IV
Technology Use in Tax Compliance and Tax Planning

CHAPTER 6
The Role of Automation in Improving Indirect Tax Reporting and Compliance

Stephen W. James and Erik van der Hoeven

§6.01 INTRODUCTION

How can companies manage their increasingly complex global indirect tax compliance requirements in a cost effective way? Every company must manage its tax compliance and associated administrative requirements to avoid challenging tax audits and severe penalties. Governments coerce compliance through frequent checks, substantive audits, penalties and even prosecution. Companies often deploy tax automation solutions to help them with complex and constantly changing tax rules and regulations and to reduce the costs of tax compliance. Tax automation solutions, typically, maintain indirect tax rates and rules within a company's financial system, and can be essential to eliminating costly mistakes and improving efficiencies.

Large businesses are generally better positioned to manage their indirect tax demands by developing strong internal control systems and procedures, and using their existing computer systems to manage tax. Small to medium enterprises, on the other hand, frequently have the same compliance complexities, yet they face a world of difficulties in understanding and complying with tax requirements. Their ability to create, record and maintain reliable records, and make them available for audit by tax and regulatory agencies, as an integral part of their normal operations may be somewhat limited due to constraints in human resources and/or budget. Hence the development of commercially available tax software packages was an important step towards reducing their cost of compliance.

The Organization for Economic Co-Operation and Development (OECD) has issued a guidance note that highlights the issues that need to be managed and illustrates the tax reporting and filing, tax auditing and corporate governance challenges for all corporations. Broadly, the OECD paper discusses best practices and standards in tax software, tax reporting and filing with a view to the deployment of technology being the means to archive cost reductions for both businesses and tax administrations alike. The OECD gives some guidance on how the principles formulated in the paper could be implemented in practice, and also provides some insight into how software developers and revenue authorities operate to develop and support certain standards. Many of the underling points that drove the discussions in the OECD paper are similar to those that have driven companies to look to technology as a means to help optimize their indirect tax compliance processes.[1]

From a business systems perspective, accounting systems are typically designed to produce only basic information for tax reporting purposes, and not to manage tax from procurement to payment. Hence, accounting software can produce reports that can be used for tax reporting purposes, but cannot generate actual returns or allow for electronic filing. Auditors are now also faced with an increasing verification challenge whereby advances in technology and a growing number of legacy operating systems, data formats, backup and file retention options make their task increasingly complex. Finally, corporate governance is becoming an even more important area of scrutiny as governments worldwide continue to demonstrate firm resolve to increase corporate responsibility and accountability through legislation such as the Sarbanes-Oxley Act 2002 in the US and many similar legislative requirements in other countries.

As a consequence of the global expansion of business and the concurrent global shift to indirect taxation, companies of every size from Fortune 100 to the Mid-market must now comply with increasingly complex regulations, and report their tax positions to a range of authorities globally—or face financial, reputational and trading consequences if they come up short.

Businesses are also facing significantly increased rate and rule volatility as countries and their tax administrations seek ways to address their issues with revenue and budget shortfalls. In addition, tax authorities conduct more targeted, more frequent, more intrusive and more coordinated audits—quite simply as little more than a reaction to the need to raise revenues and collect on those revenues as quickly as is reasonably possible.

In response, there are some prudent steps that business can take to help to mitigate these risks. In the context of this chapter, those steps typically center on tax technology and how technology can be leveraged to support processes and controls that themselves underpin tax. Broadly, technology can be leveraged to help to:

- Automate the indirect tax determination and calculation
- Automate the indirect tax compliance processes
- Build better tax planning and audit defense capabilities

The purpose of this chapter is to provide some insight into the typical tax challenges faced by today's tax leaders and how those challenges are being addressed by the smart deployment of technology. We begin in section 6.02 by discussing the regulatory

---

1. OECD, Centre for Tax Policy and Administration, *Guidance on Tax Compliance for Business and Accounting Software*, May 2005.

environment in which tax departments presently have to operate. In section 6.03 we will explain the function of the most important tools available in the area of indirect tax automation. Enterprise Resource Planning (ERP), bolt-on tax engines, reporting solutions and a move to hosted or SaaS-based software solutions are all key components that impact the way businesses manage their indirect tax obligations worldwide.

In our experience, most tax departments will derive significant benefits from the introduction of automation in the compliance processes. We will also explain where we, typically, see greatest value in an investment in tax technology and ways to help to maximize a business' return on any investment in tax and technology. In section 6.04 we describe the process of a typical tax automation implementation project. Next, in section 6.05, we summarize the benefits and value a company may expect to derive from the application of tax technology. The role of technology in supporting tax strategy is discussed in section 6.06, and our conclusions and predictions for future technology development in the area of taxation are found in section 6.07.

§6.02 REGULATORY ENVIRONMENT

From a global perspective, the top two fiscal issues that countries are struggling with are fraud and budget and revenue shortfalls. The so-called revenue gap, i.e., the difference between actual tax revenue collected and the predicted tax revenue that should have been collected based on recasting historic economic data is substantial for many countries. Numbers vary, but according to some reports the gap in the EU alone is significant with reports of more than USD 300 billion suggested in some quarters as the annual cost of VAT fraud. Overall, revenues are shrinking as result of a prolonged economic recession. VAT fraud, in particular the so-called missing-trader or carousel fraud schemes,[2] have become the territory of organized crime and cost governments billions in revenue.[3]

All of this is translating into a focus on increased audit and a renewed focus on the quality and capability of business systems, particularly those that handle tax. These systems are, of course, increasingly required to support and manage highly complex business models and value chains and the increasingly complex tax requirements— in terms of rates, rules and reporting— that arise. What's more, the tax function is under increasing pressure to deliver better results and to deliver these results with fewer resources. Tax leaders are, simply put, being asked to do more with less.

---

2. See on this subject, Robert F. van Brederode, Third Party Risks and Liabilities in Case of VAT Fraud in the EU, 2008 (34) 1 *International Tax Journal*: 31-39. For specific technology driven solutions to mitigate fraud, see Richard T. Ainsworth ' VAT Fraud and Technological Solutions, in: The VAT Reader: What a Federal Consumption Tax Would Mean for America, Washington, DC, (Tax Analysts), 2011: 204-223.
3. Some estimates place the cost of VAT fraud within the EU alone at USD 340 billion per annum: John Norregaard and Tehmina S. Khan, "Tax Policy: Recent Trends and Coming Challenges," 39 (IMF, Working Paper 274, 2007). According to a recent PricewaterhouseCoopers report an estimated 11 percent of VAT revenue is lost annually through fraud, which equates to in the region of €100 billion: Shifting the balance from direct to indirect tax, 2011 at 15. (http://www.pwc.com/gx/en/tax/indirect-taxes/shifting-balance.jhtml)

In early 2011 the SEC fined Hudson Highland Group Inc. for lacking the internal controls to properly collect and remit nearly USD 4 million in sales taxes.[4] The SEC ruled that From March 2003 to January 2007, Hudson's North America segment (HNA) "failed to consistently comply with tax laws that required it to collect sales taxes from its customers, and to remit them to the taxing jurisdictions. The reason for these failures is that HNA did not have accounting software capable of calculating the amounts of sales taxes owed."[5]

Section 13(b)(2)(B) of the Exchange Act[6] requires all reporting companies to devise and maintain a system of internal accounting controls sufficient to provide reasonable assurances that transactions are recorded as necessary to permit preparation of financial statements in accordance with generally accepted accounting principles. In addition to nearly USD 4 million in tax that had to be paid out from the company's own reserves, the SEC levied a significant fine. The case centered around the Exchange Act expectations that controls in place are reasonable and effective.

This is the first of what could be many such penalties. The bottom line is that these issues are avoidable— companies applying the right controls and focusing on the right things are far less susceptible to penalties of this kind.

§6.03 TECHNOLOGICAL SOLUTIONS

What the previous section also demonstrates is that the challenges faced by tax leaders are, to one degree or another, almost universally applicable and, again to one degree or another, systemic in nature. These are repeatable, process-driven challenges that are, in our experience, best addressed through technology.

Broadly, it is driving to how companies choose to manage risks and there are a number of ways in which companies can and should manage and mitigate them. Obviously, companies can plan to avoid the risk. Here, it may be possible to change how a transaction is structured so that it creates a different or lower risk tax outcome. It may be possible to offload some of the risk to external advisors or through the use of outsourcing. It may furthermore be possible to reduce the financial impact of these risks by making provision for the taxes that may or may not become due.

Finally, we come to what typically drives greatest benefit and return on that investment— the reduction of risk through a decrease in error due to the automation of key parts of the tax and business processes. Businesses can reduce the impact of people, particularly non-tax professionals, and leverage data that are needed anyway for other business purposes, thus reducing cost, effort and risk. This is what tax automation does best; adding enormous value.

---

4. Securities and Exchange Commission, Release No. 34-63688, File No. 3-14182, January 10, 2011.
5. *Supra*, Summary.
6. Securities Exchange Act of 1934 (Pub.L. 73-291, 48 Stat. 881, enacted June 6, 1934, codified at 15 U.S.C. §§ 78a et seq.)

Below we will explain some of the key concepts and technologies that should be high on the agenda as companies are considering how best to address their tax and business needs.

[A] Capabilities and Limitations of Enterprise Resource Planning Systems (ERP)

The proliferation of all encompassing software solutions for managing major business processes such as procure-to-pay, order-to-cash and all other associated processes started in the 1990s. Today, a majority of companies use software to manage most aspects of their business, with solutions ranging from single-user financial software to integrated computer-based systems to manage tangible assets, financial resources, materials, people assets, and more. This all encompassing software is referred to as Enterprise Resource Planning, or ERP, and its leading suppliers include SAP, Oracle, Microsoft Dynamics, PeopleSoft, Baan, JD Edwards, Epicor and Sage.

ERP applications exist to facilitate the flow of information across all business functions inside the boundaries of the organization and to manage connections to outside stakeholders. An ERP system is typically built on a centralized database and utilizes a common computing platform in order to consolidate all business operations into a uniform enterprise-wide environment. ERP software is traditionally installed inside the firewall of a company on its own servers. Since the early 2000s, ERP software is also becoming web-based, allowing both employees and external resources (such as suppliers and customers) real-time access to system's data through the internet.

Companies around the world use ERP software to support almost all aspects of business, including sales, marketing, delivery, billing, production, inventory management, quality management, human resource management, and taxes.

The ERP system is core to the effective and efficient management of most businesses with operations of even moderate scope and scale. Simply put, the ERP are designed to enable and facilitate the flow of information and data across the organization, supporting a whole host of key business functions such as finance, sales, manufacturing and SG&A etc. The relevance from a tax perspective is that many if not most of the transactions and business functions that have tax consequences, for example sales, procurement, inventory and logistics— are powered and enabled by the ERP application.

The challenge is that ERP systems typically come with little to nothing in the way of tax functionality and tax content out of the box.

ERP systems typically require configuration and customization to handle indirect taxes and do not deliver rate and rule content. Any rate and rule content will typically need to be created and will require ongoing maintenance. The bottom line is that no mainstream ERP system is designed to specifically address the needs of tax, and none deliver robust out-of-the-box indirect tax functionality.

The ERP paradigm for tax is, broadly, that tax decisions are driven by the selection of a tax code at the point a transaction is processed. While some ERP systems can support the semi-automated selection of tax codes based on business rules and transaction attributes, the reality is that manual input, in the form of manual code selection, is driving taxability at most organizations. Furthermore, in many cases it is non-tax professionals in the form of AR and AP resources which are the ones making these key decisions with clear tax impact.

A major challenge with the ERP paradigm for tax is that the tax function will typically be heavily reliant on IT and other non-tax resources to implement change. In this regard, change might speak to changes in the underlying business that require the ERP tax content and/or business logic to be updated, or it might be changes in the legislative framework or judicial interpretation that might require changes to be made in the ERP.

The ERP systems that are most commonly in use at the larger organizations are Oracle and SAP. At the time of writing, Oracle's current mainstream release is Oracle R12 and with this release Oracle has added some key tax functionality. In the past, in Oracle 10 and Oracle 11i, the application provided some global indirect tax functionality but it was generally limited to AR and to a lesser extent AP. Oracle R12 introduces eBusiness Tax, a new Oracle module that is intended to effectively be Oracle's tax engine. eBTax is available to most Oracle modules, so now, for the first time, the tax module can be used to support some of the other key Oracle modules including those that support fixed assets should a business so desire. In the past, this would have been very difficult, primarily due to the focus on AR and AP.

SAP has generally been a little more advanced than Oracle in terms of VAT functionality, though its own solution is not without its complexities and challenges. It is again broadly a VAT code and condition-based solution where, essentially, logic or manual processes are created that look at a transaction, review the data, find a match and choose a tax code.

In both systems, there are two major issues. Firstly— there's no content. Both solutions provide a framework for tax, but no content to leverage that framework. Typically, this content will need to be created, and from then on it will need to be maintained. As a matter of interest, Oracle and SAP are both set up to expect a tax engine by default for North America, something we cover later in this chapter, but both default to their native tax solution for the rest of the world. We will cover support for the global market later in the chapter too.

In terms of the other major ERP applications, you will find fewer deployments (particularly global deployments) with PeopleSoft (now part of Oracle), JD Edwards and Baan.

Oracle

- 10.7/11i and prior supported sales and purchase taxation
- R12 and eBusiness Tax now supports more Oracle modules
- eBTax is Oracle's hybrid ERP/Engine tax solution

Chapter 6: The Role of Automation in Improving Indirect Tax Reporting   §6.03[A]

SAP

- SAP R/3, ECC 6.0/7.0 has relatively advanced VAT functionality
- SAP uses tax code and condition-based solution for VAT
- SAP has a suite of Business Intelligence tools to help facilitate reporting

Let us look at why this is important. Worldwide, more than 15,000 taxing authorities levy indirect taxes on the production, movement or sale of goods and services. In the USA, for example, just for sales and use taxes there are more than 7,500 taxing authorities.[7] All these taxes are levied at different rates, and apply varying rules to determine taxability and tax jurisdiction. Often, elements like the exact place where a good or service is enjoyed, or how a product will be used, determine the applicability of a tax. Moreover, hundreds if not thousands of rules and rate changes occur worldwide each month.

ERP penetration is about 50-60% in the USA and lower in the rest of the world. That means that about half of all companies are managing indirect taxes using other tools. The most common tools are accounting packages or office tools, like MS Excel. For most companies that would be sufficient, especially when they buy and sell in one region and are familiar with the local tax rates and rules. It becomes much more challenging for companies that change vendors frequently, acquire new customers and do businesses in larger regions where different rates exist.

For example, in the city of Denver, Colorado State, the state sales tax rate is 2.90%, the county sales tax rate is 0.25% and the city sales tax rate is 3.75%. On top of that, there is a football stadium district tax of 0.10%, a regional transport district sales tax of 1.00% and scientific and cultural facilities district sales tax of 0.10%. In total, 8.1% sales tax applies for six different tax authorities. This example illustrates the complexity of having to manage and report the complexities around combined sales tax rates.

Most ERP suppliers provide a standard set up to manage taxes within their system. For example, SAP offers three standard methods to compute taxes on sales and purchases or sales and use taxes: SAP Non-Jurisdictional Taxes on Sales & Purchases, SAP Jurisdictional Sales & Use Taxes, and SAP Tax Interface to third-party external tax packages.

Nevertheless, there can be shortcomings to managing indirect taxes in an ERP system. For one, ERP systems are complex and thus even larger companies that have installed ERP systems do not always utilize the innate ERP capabilities of managing indirect taxes to their full advantage. For another, a company needs to put significant effort into researching tax changes as they take place, which could impact their business. The rates and rules need to be maintained in the ERP system and staff needs to be instructed on using the correct tax treatments. In the USA in particular, up to six different taxes are possible in parallel, resulting in tax scenarios with many thousands of tax jurisdictions for which time-dependent tax percentage rates must be

---

7. See Robert F. van Brederode, *Systems of General Sales Taxation: Theory, Policy and Practice* (Kluwer, Series on International Taxation 33, August 2009), Chapter 6: 71.

maintained, taxes paid periodically, and to which tax revenues must be reported. These factors make maintaining tax percentage rates manually or communicating with tax authorities very time-consuming since formats can vary from authority to authority.

However, purpose-built tax management software products are available that are tailored to these scenarios, allowing taxes to be calculated, reported, and paid where such special rules apply. In addition, the companies that provide these products keep their customers up-to-date concerning changes to percentage rates and calculation types, thereby reducing manual maintenance time considerably, especially for companies that offer direct sales or have a large number of organizational units in different locations. For these reasons, most ERP suppliers support generic interfaces to such third-party, purpose-built tax engines, enabling them to be "bolted on" to the ERP system. These interfaces enable all the necessary data flows at the appropriate points, thus supporting seamless integration.

For this reason, companies are increasingly opting to use these bolt-on tax engines, linked to their ERP systems, to maintain and apply the correct tax rates at the correct time.

[B] Tax Solutions beyond the ERP

In addition to the ERP-based tax solutions, there are a number of tax solutions and services that seek to provide more comprehensive capabilities in relation to indirect tax. The common theme is that these solutions and services seek to replace the native functionality in the ERP and replace it with more robust capabilities and, equally importantly, content in the form of tax rates, tax rules and tax logic.

[1] Tax Engines

The tax engine is the most robust and most capable of the solutions that seek to replace the native tax functionality in the ERP. The power of the tax engine is that it lets the organization make the most of its existing business data, data it would require in any event, and leverage that existing business and transaction data to drive predictable and repeatable indirect tax determination, calculation and reporting. There exist a fundamental philosophical difference between an engine and the ERP in that the engine is built on the vision that tax logic and decisions should not be hard-coded or pre-assigned as it must be in an ERP environment.

The engine, broadly, exists to take data elements that describe a transaction and use those data elements to make a tax decision in much the way as a tax professional would interpret the facts of a transaction when determining its treatment. This data might, *inter alia*, comprise product information, accounting data, shipping data, pricing information, tax registration data and terms of business. The engine, in essence, aggregates this data to work out what is being supplied and the jurisdictions entitled to tax it. The engine then leverages its rate and rule data to perform the calculations.

One of the important things to understand is that the engine must be integrated with the ERP system to enable the data residing in the ERP to be passed to the engine. These integrations are, in the main, standard and proven integrations that leverage the standard integration points provided by the ERP vendors.

The engines will drive tax decisions that are more easily managed, archived and audited and, importantly, will reduce the influence of users on the final tax outcome. Unlike in ERP taxation, users typically do not select a tax code when a transaction is being processed. Simply put, the engines seek to take this requirement away in the hope that this particular automation paradigm can deliver greater reliability and consistency in the tax processes and the repeatable application of tax policies. The engines also seek to help to build a low maintenance solution, one that has greater flexibility to handle future tax and business challenges when compared to an ERP tax solution.

There are three, main, vendors of global tax solutions. Sabrix[8] from Thomson Reuters, Taxware[9] from ADP and Vertex[10] from Vertex, Inc.

Arguably, the most important benefit of using a tax engine is the content that is provided and maintained by the software vendors. The content that is supported falls into two main categories. First, the software needs to be able to discern what "place of supply" logic applies to a certain transaction. For example, when goods are sold from a UK-based company to a German-based company, is VAT applicable and if so, which VAT is charged, the UK or German one? Most tax engines are engineered to take many elements into account to make the appropriate determination. In this example, a valid German VAT registration and proof that the goods are shipped from the UK to Germany, can determine the appropriate taxability of the transaction.

The other important area of content is the rates and rules. Tax engines need to be able to support all standard, reduced and exempt rates. Moreover, they need to have content that can drive the correct taxability for specific goods and services as well. In addition, some countries (especially in the European Union) require special messages on invoices that will have to be provided by the tax engine. In our example of goods being sold from the UK to Germany, the result could be that the UK seller should not charge UK VAT, because the German buyer must self-assess German VAT. The seller must place a special message on the invoice with a reference to the appropriate legislation to indicate this treatment.

ERP systems can to some degree handle tax requirements, but they were never specifically designed for managing tax. Therefore, these systems are not always best equipped to deal with more complex tax transactions. It is not uncommon for multiple companies to take part in a chain of business transactions. One company could receive the order, a third party may produce the product and ship it directly to the buyer. These so-called "triangular" transactions can create serious challenges for ERP systems to manage. Tax engines, on the other hand, should be able to process all the transaction

---

8. http://onesource.thomsonreuters.com/solutions/indirect-tax/
9. http://www.adp.com/solutions/employer-services/sales-and-use-tax/large-business/taxware-enterprise.aspx
10. http://www.vertexinc.com/solutions/indirect/

data and determine that a triangulation is in place and assess the appropriate tax results. A tax engine is able to determine tax based on all the elements of a given transaction. Often more than 250 elements can be considered, including address, customer, products, pricing, tax jurisdictions, customer/vendor taxability and much more. In contrast, an ERP system only considers the tax code that is passed manually to it. Therefore, the premise is that these transactions can be easier set up in a tax engine than in an ERP.

[2] Hosted Tax Solutions

Hosted tax solutions differ from tax engines in a number of ways, but the most fundamental are (a) that there is no real time integration and (b) that the engine is not installed on the company's own network and behind the firewall. Rather than make tax decisions via a direct connection to the ERP, hosted solutions are designed to take data via internet-based tax calls. They tend to be lower cost, tend to be easy to integrate but also tend not to have the full range of capabilities that would be delivered by an on-premise tax engine. Some organizations will also be less comfortable with exposing data to the internet (an inevitable consequence of the solution being hosted).

Some of the tax engine vendors listed previously also offer hosted solutions. Other key hosted solution providers include: Avalara[11] and Speedtax.[12]

[3] Reporting Solutions

One of the most often overlooked aspects of managing the tax function is the requirement to ultimately have to report tax data on various local and federal returns and other similar filings. Returns packages for the US have long been with us and are broadly well known and well understood. In recent years, returns packages for the international markets have become available and these solutions are increasingly able to facilitate VAT reporting and compliance. Some like iVAT[13] and Abacus[14] are multi-country solutions, some such as Mastersaf[15] and Syncro[16] are point solutions focusing on specific countries— in this case Brazil.

§6.04 THE IMPLEMENTATION PROCESS

Most tax technology projects tend to follow a typical process with certain key phases that together comprise the necessary work to implement the solution. The key phases of a tax technology project are, broadly, sequential though some work will span

---

11. http://www.avalara.com/.
12. http://www.cchgroup.com/webapp/wcs/stores/servlet/content_LP_speedtax.
13. http://www.vatat.com/index.php?page=ivat-reporting&hl=en_US.
14. http://thomsonreuters.com/products_services/taxacct/united_kingdom/onesource_indirect_tax/compliance/.
15. http://www.mastersaf.com.br/.
16. http://www.synchro.com.br/site/.

multiple phases and some may be performed in parallel. The proper implementation process is key to the success of the overall project. It is here that projects succeed or fail and, in our opinion, the chances of success are maximized through the application of a proven and tested methodology.

The first phase of any project focuses on **requirements**— here, the tax, business and functional requirements need to be defined that will be supported by the tax engine implementation. This will comprise a clear understanding of the business and the extent to which taxes impact the supply and value chains. As the company works with its suppliers, customers and internal partners, a number of critical points will require the delivery of a tax decision. In the requirements phase, a clear understanding should be acquired of where these points are and what should drive the decisions.

The second phase is **design and blueprint**— this is where the objective of the team is to design an efficient and effective solution, one that can be maintained by tax and which is scalable to support future growth. Design, in this context, means the way that the specific features of the ERP and tax solution will be utilized and leveraged to support the requirements identified in phase one. Paradoxically, a complex design is very easy to build. It takes much more effort and skill to be able to create a simple and elegant design, but the benefit in doing so is that the solution will be much more manageable and much more scalable as a consequence. The complex designs fall down as they tend to focus on individual transactions and, as such, there tend to be a lot of configurations focused at a very low level. The elegant designs are much more sophisticated and work at a higher level, utilizing the power and logic embedded within the tax solution in a far more effective way.

The next phase is **build** where the objective is to create the engine or solution configurations and help support the ERP integration. As noted in the design discussion, this should be done using proven software engineering principles (leveraging proven libraries) to create a very elegant and manageable solution where effective and efficient configurations make the most of the engine's capabilities, features and benefits.

In the **testing** phase the aim is to make sure the solution delivers the right tax for the right reasons. There are typically three parts to the solution testing. Integration testing seeks to ensure that there is robust and bi-directional connectivity between the ERP and the tax solution. Unit testing is where the solution is tested for defects using what will ultimately be the release candidate configurations. User acceptance testing is where actual users simulate the production environment by running transactions from ERP and tracing the results. Testing is an absolutely critical part of the solution and must be carefully planned and adequately implemented if the levels of confidence in the solution are to be where the organization needs them to be prior to cutover.

In the **deployment** phase or **cutover,** the configurations are deployed into the production environment and are made live. This can be a technically challenging phase, especially when the project is working with a Brownfield implementation (where the host ERP system is already in production) rather than with a Greenfield implementation (where the host ERP is being deployed concurrently with the tax solution).

It is important to understand that these solutions are complex and that the process to deploy them can be complex as well. In our experience, a number of specific issues may arise on these projects and they are listed here to provide a note of caution. One of the most prevalent is underestimating the time needed to successfully deploy a complex and often global tax solution. Others include a failure to build a robust and manageable project plan, a failure to set project expectations at the right level and a failure to adequately gather requirements at the outset. Building on these, other challenges have included a failure to adequately test the solution and, crucially, a failure to get the right level of buy-in from stakeholders and users. This is a really critical point, projects will not be successful unless all stakeholders are aligned and driving towards the same preferred outcome.

A final point, and worthy of discussion on its own, is a failure to build a solution that can be managed and supported. It is important to understand that it is relatively easy to build a complex and unwieldy solution and much harder to build a simple, focused and elegant solution design. The difference is that the latter is very supportable, the former is not.

To put things into context, it's very easy to forget the scope and scale of indirect taxes. A typical USD 5 billion organization might, for example, ultimately drive a USD 150 million income tax liability but that same organization might process and manage over USD 1.5 billion in indirect taxes. Typically, organizations tend to have more income than indirect tax resources.

One overarching question any organization should contemplate on is whether it is focusing on what is urgent and not on what may be more important. Many organizations tend to focus on lower value, less strategic activities such as compliance, in large part due to the resource constraints that were discussed earlier.

§6.05 BENEFITS AND VALUE OF TAX TECHNOLOGY APPLICATION

When considering an investment in tax technology, one of the most important determinations that needs to be made is whether that investment is likely to deliver the anticipated features and benefits.

In our judgment, one of the most critical deliverables of tax technology is the ability to better and more proactively manage change. When the tax decision and tax compliance process are data-driven rather than products of the choices made by people, the key benefit is that changes in rules, rates, product taxability and the like can be supported as a natural product of the underlying tax content being kept current. Content updates for the major tax solutions tend to be applied monthly with special out-of-cycle updates typically being pushed out to users when critical changes need to be reflected before the next regularly content update would ordinarily have been delivered.

One benefit that can add tremendous value for the modern tax department is the ability to model and test global scenarios in a tax workbench environment which leverages the capability of the tax solution in a way that lets users run complex tax scenarios to predict the outcome and model the impact of changes to those scenarios. Most of the modern tax engines offer this facility and it is something that can deliver tremendous value to those who have the ability to use it. Not only can the workbench-type capabilities be used to confirm taxability, etc. for current business scenarios, the solution can deliver taxability predictions for hypothetical scenarios and enable the tax department to plan with greater confidence.

We have also seen companies drive significant value through the fact that most tax solutions provide a single audit database which can greatly streamline and facilitate the global tax data gathering processes which underpin compliance and reporting. In our experience, tax departments tend to struggle with this and typically tend to spend a disproportionate amount of time in aggregating and managing data as they prepare to file returns and other statutory disclosures at the end of each reporting period. The typical problem is that the data needed to drive the compliance processes might reside in multiple systems and might be in different formats and of different provenance. Business intelligence tools such as tax data warehouses seek to address this by providing one repository for tax data and a way to normalize data that might have been sourced from different systems and thus have different characteristics and attributes. The normalization process seeks to bring data together and let it be described in a consistent way that makes it more usable and more accessible to users and thus of greater value in the compliance process.

Tax engine solutions also aid as transactions processed by the engine are archived in a single database together with a clear audit trail showing the data used to support the tax decision and the specific rules and logic that were triggered to support the decision. This drives a much greater ability for the business to support those decisions upon audit and, of course, drive better audit management capabilities through the ability to be more proactive and less reactive in how data is screened and made available to the auditor.

Consequently, it's easy to see how tax technology solutions, when deployed in support of the end-to-end indirect tax processes can add value in terms of their ability to support local reporting, both management and statutory,

One final advantage is that tax solutions tend to benefit from the ability to deploy business-specific global configurations in once place. Today, many organizations find they have had to deploy tax logic and tax content in many different systems. A company with, say, five different billing systems, a procurement system and an inventor system may find it needed to maintain country-specific tax logic and tax rules in all seven systems, which can add cost and complexity. Modern tax solutions, which in some cases can support the capability for one tax system to simultaneously support several different systems, let all of the tax logic and content reside in one place. This is a key feature that reduces the need to manage and maintain redundant data in multiple systems.

§6.06 THE ROLE OF TECHNOLOGY IN SUPPORTING TAX STRATEGY

Tax strategy can be defined as *"the intelligent allocation of tax resources, tax information and tax knowledge to deliver lasting operational benefits and competitive advantage."* An effective tax strategy will support and embrace the importance of indirect taxes and give it equal priority to the direct tax function. It must also align tax, and those who define the success of the tax function, with key organizational goals and objectives. Finally, it must encompass key aspects of the entire end-to-end tax process, such as effective rate management, effective tax cashflow management, reporting (financial, statutory, management), supply chain, indirect taxes, audit management, tax compliance, tax structuring and tax planning, etc.

We have discussed the benefits of indirect tax technology and we consider it key to the effective deployment and management of an organization's tax strategy. Technology enables true insight into how tax is supporting the business and it helps to bring together the various data that enables tax to add greatest operational and business value to the organization.

§6.07 CONCLUSIONS AND PREDICTIONS

Looking forward, we expect to see evolution and not revolution in the tax automation space. Our position is that the last major revolution happened in the early part of the last decade when tax solutions, for the first time, became capable of supporting the global operations of a complex multinational organization. While ERP systems had some capabilities in this regard, those capabilities were neither robust enough nor scalable enough to truly meet the needs of more complex tax organization.

We are predicting that a major part of that continued evolution will be focused on the demands placed by the more complex countries. The BRIC countries in particular are likely to be an area that will receive a lot of attention from the technology providers, given that these countries are increasingly on the road map for many global companies. Furthermore, these are countries that impose significant burdens and introduce significant complexity into the calculation and determination process and, furthermore, have complex compliance requirements such as the SPED in Brazil or the Golden Tax System in China. These again are challenges where technology can and will play a key role in their resolution.

We also anticipate an increasing trend in that tax administrations are placing their own focus on reporting and compliance and the quality and capability of the business systems that underpin them. In the Netherlands, for example, the Tax Control Framework is a mechanism by which a company can submit its tax and business systems for review and, broadly, expect a lighter touch upon audit and in terms of ongoing interest from the authorities in relation to its tax compliance requirements. This systems-based review process is, we believe, a preview of what we can expect to see from other countries as we look forward. The thinking is that a company with strong controls and a willingness to submit those controls for review is, more likely than not, a company which is going to deliver a greater level of compliance. Not unreasonably, the authorities are going to spend their time focusing on the organizations which have shown less willingness to adopt a systems-based control program and, furthermore, submit those controls for review and scrutiny.

We are predicting that businesses will also be demanding much greater levels of integration than we see today in terms of how tax technologies work together. Today, many parts of the indirect tax end-to-end process can be supported by technology but more often than not there are different solutions powering different parts of the process. A company may handle tax calculations for the US with a tax engine, handle calculations for their global operations in their ERP, handle reporting using business intelligence tools, and use a number of different returns applications or outsourcing to handle compliance. In future, we predict greater convergence and the emergence of tools that support more of the end-to-end process, or suites of tools that were built to work together to create that end-to-end solution.

An additional feature and benefit that we believe will be key is the ability for tax and business leaders to leverage technology to drive greater visibility into the current state of their tax processes and the extent to which their global operations are current in terms of their filing requirements. Dashboard-type solutions exist today which can aggregate data and present it to the tax leader in a way which is easy to understand and easy to interpret at a glance, providing comfort and/or early warning depending upon how the global operations perform in relation to their filing and payment requirements.

Finally, we predict that the requirement to electronically report tax data in real time, or close to real time, to the tax authorities, broadly mirroring to some extent the positions we see today in China and Brazil, is likely to become more prevalent. As counties struggle with revenue and budget shortfalls, there will continue to be pressure on the VAT systems to deliver revenue as close as possible to the time the business transactions that drove it were performed.

To conclude, we believe that technology has become a vital component of an effective tax strategy and, when we consider the typical challenges that today's tax leaders are experiencing, we are yet to find a better way to address them than technology.

§6.08 BIBLIOGRAPHY

Ainsworth, Richard T. 'VAT Fraud and Technological Solutions,' in: *The VAT Reader: What a Federal Consumption Tax Would Mean for America*, Washington, DC, (Tax Analysts), 2011: 204-223.

van Brederode, Robert F. Third Party Risks and Liabilities in Case of VAT Fraud in the EU, 2008 (34) 1 *International Tax Journal*: 31-39.

van Brederode, Robert F. *Systems of General Sales Taxation: Theory, Policy and Practice* (Kluwer, Series on International Taxation 33, August 2009), Chapter 6: 71.

Houtzager, Mark. 'Technology: The Crucial Pillar of VAT Implementation,' in *The VAT Reader: What a Federal Consumption Tax Would Mean for America*, Washington, DC, (Tax Analysts), 2011: 224-229.

Norregaard, John & Tehmina S. Khan, "Tax Policy: Recent Trends and Coming Challenges," 39 (IMF, Working Paper 274, 2007).

OECD, Centre for Tax Policy and Administration, *Guidance on Tax Compliance for Business and Accounting Software*, May 2005.

PricewaterhouseCoopers, *Shifting the balance from direct to indirect tax*, 2011 at 15. http://www.pwc.com/gx/en/tax/indirect-taxes/shifting-balance.jhtml

Walsh, Chris. "How You Can Manage Taxes in a Hyperregulatory World," *International Tax Review*, May 2010.

CHAPTER 7
Automation Technology in Special Customs Programs and Preferences
William M. Methenitis[*]

§7.01 INTRODUCTION

A customs duty is a tax charged at the time of importation of goods arriving into a particular country or area from outside that country or area. Customs duty is assessed on the imported good, usually at an ad valorem tax rate (i.e., a fixed percentage of the value of the imported good). Duty rates are product specific, with rates often varying significantly depending on the type of product imported. Rates may also be dependent on the country of origin of the imported product. Consequently, to accurately compute the amount of customs duty due at importation, it is necessary to know the customs classification of the product (a numeric code that is partially harmonized in all WTO countries), the country of origin of the product, and the correct customs value of the product, which may or may not be the invoiced price.

    The amount and timing of the payment of customs duties is also impacted by special programs and preferences. Special programs are country specific duty concessions provided to encourage activity in the country. A number of countries, for example, offer special programs that eliminate duties on imported products that are further manufactured for export. Preference programs provide preferential duty rates provided that specific criteria are met. Free trade agreements, such as North American Free Trade Agreement (NAFTA) and Association of Southeast Asian Nations (ASEAN), are preference programs that provide a lower duty rate for items produced within an identified region. Special programs and preferences are quite pervasive; in fact, many businesses avoid more customs duties by utilizing special programs and preferences

---

[*] The author gratefully acknowledges the assistance of James Grogan of Ernst & Young, LLP in the preparation of this chapter.

than they actually pay. It is important to note that these programs are almost universally conditional rather than automatic; that is, the benefit accrues to the importer only if particular conditions are met and records are kept.

In addition to the direct financial impact of customs duties, there are also important operational consequences of the customs function. Correct customs filings are required for products to move into a country. Incorrect or incomplete filings can therefore add hours— sometimes days— to the supply chain, which can significantly affect the importer's operations. Inaccurate filings can lead to penalties, and in extreme cases, the seizure of goods.

Because customs duties are transactional taxes, assessed at the time of importation, every separate importation of products requires a separate filing. This makes the customs function very data intensive and very process driven.

Against this background, and with international trade of increasing importance to many businesses, automation has become an essential part of conducting cross-border trade and managing the data necessary to meet compliance obligations. Automation has also become increasingly important to businesses looking to take advantage of special customs programs and preferences. Some of the automation has been government mandated. More frequently, however, automation has developed to help business take advantage of specific applications of special customs programs that might not have been achievable absent enabling technology.

This chapter will explore several different aspects of automation as it relates to special customs programs and preferences. Topics include government-mandated automation programs, automated solutions developed specifically to achieve customs duty savings, and the use of automated customs tools to facilitate performance optimization.

§7.02 GOVERNMENT-MANDATED AUTOMATION

Government and business have a long history of cooperating on electronic data transmissions related to the transportation of goods. Many jurisdictions (e.g., United States, Mexico, certain European Union countries) have also adopted electronic filing of customs data, and in some cases require advance electronic notification of shipments to assist with cargo security and import risk assessments. Standard use of Electronic Data Interchange (EDI) for governments and businesses began in the early 1980s, and has improved the way government agencies and businesses share their data, eliminating the need for paper systems and data recapture, and significantly reducing the chance for errors. For example, US importers and customs brokers can manage much of their customs business (e.g., cargo e-manifests, customs entry documentation, electronic payment of customs duties, etc.) electronically through the Automated Commercial Environment (ACE) Secure Data Portal.[1]

In some cases, access to special customs programs has been conditioned on automation. In these instances, businesses may access special and important customs

---

1. http://www.cbp.gov/xp/cgov/trade/automated/modernization/.

benefits, but only if they agree to automation protocol which better enables compliance and government enforcement. Prime examples include the IMMEX program in Mexico and the RECOF program in Brazil, which are described below.

[A] Mexico: The IMMEX Program and Automated Inventory Controls

[1] Overview and History

The IMMEX[2] program is a special customs regime in Mexico which allows consignment or contract manufacturers to import raw materials, machinery and equipment into Mexico on a temporary basis. Imported raw materials used in the manufacture of exported finished products are generally exempt from import duty, as they are not destined for the Mexican domestic market. Furthermore, the importation of these materials into Mexico on a temporary basis does not give rise to value added tax. Mexican manufacturing companies qualify for the program provided they have annual foreign sales of at least USD 500,000 (or equivalent in Mexican pesos) or invoice abroad at least 10% of total export sales. IMMEX is designed to increase exports and encourage local manufacturing operations by providing customs and tax benefits to qualifying participants.

Prior to 2006, there were two temporary import programs in Mexico: PITEX[3] and the maquiladora[4] program. The primary difference between the two programs was that PITEX applied primarily to entities formed by Mexican investors which typically own all operating assets and materials, whereas the maquiladora program was designed for foreign investors who consign equipment and raw materials to the Mexican business for processing. In November 2006, the government merged the two programs into the current IMMEX program.

[2] Automated Inventory Controls

The Mexican Customs Law states that all importers, including those operating under a temporary import regime for export, such as the IMMEX program, must have an automated system for inventory control.[5] Furthermore, the IMMEX decree mandates that IMMEX users implement automated inventory controls to track inputs (temporary importations), materials used in processing and finished products in accordance with

---

2. Manufacturing Industry, Maquiladora, and Exportation Services Program (*Decreto para el Fomento de la Industria Manufacturera, Maquiladora y de Servicios de Exportacion, Diario Oficial de la Federación [D.O.] 1 de noviembre de 2006 (Mex.)*).
3. Temporary Importation Program to Produce Articles for Export (*Programa de Importación Temporal para Producir Artículos de Exportación*).
4. A maquiladora (or maquila) is a factory or manufacturing plant that imports merchandise duty-free for use in the assembly or manufacture of exported finished products.
5. Article 59-1, *Ley Aduanera* (Mexican Customs Law).

the provisions of Annex 24 of the Mexican regulations.[6] From the government's perspective, temporary imports must be accurately tracked in the inventory control system in order to monitor the deferred (or eliminated) duty awarded with this program.

The automation requirement was enacted as part of the 2006 redefinition of the IMMEX program. Prior to this requirement, a lack of inventory automation, especially for complex importers, made customs audits very time consuming and difficult to manage, for both the government and the importer. The automation requirement has significantly changed the approach of the auditors, who now look to the system first. Importers without a compliant automated system can be directly sanctioned by the customs authorities. Those that do have appropriate systems find much of the audit focused on testing the effectiveness of the system.

While there have been automated inventory systems for maquiladora and PITEX companies for many years, the mandate for inventory automation has both increased the number of software providers and led to significant enhancements in system capability. Many systems today are web-based and have the capability to control multiple maquiladora facilities from a single location. As most IMMEX operations engage in frequent cross-border trade with the United States, some service providers have designed their software to work in conjunction with US software applications, allowing companies to oversee both US and Mexico operations from one application.[7]

While the Mexican government's interest in mandating an automated system was based on a desire to more effectively audit compliance with special program requirements, because the IMMEX program is so widely used in Mexico, the automation requirement has had the effect of causing significant upgrades in automated customs controls for many of Mexico's largest importers. Companies often report increased operating efficiencies and enhanced compliance as a result of the efforts to meet the government-mandated automation requirements.

[B]   RECOF in Brazil

[1]   Overview and History

Brazil's RECOF[8] program is a special customs regime that allows Brazilian companies in the automotive, aeronautics, information technology, telecommunications, and semi-conductor and high technology industries to suspend import tax payments (e.g., II[9], IPI[10], PIS/COFINS[11]) on raw materials and components (goods to be submitted to

---

6. *Anexo 24 de las Reglas de Caracter General en Materia de Comercio Exterior para 2009* (Published May 12, 2009). Annex IV of the IMMEX Decree also authorizes a simplified version of modules III and IV of Annex 24.
7. See, e.g., *Leverage Mexican and US Trade Opportunities with Integration Point Mexican Trade.* Integration Point, Inc. 2009. http://www.integrationpoint.com/products/brochures/IP_MexicanTrade_ProductBrochure_2009.pdf.
8. *Regime Aduaneiro Especial de Entreposto Industrial sob Controle Informatizado.* Regulated by the Normative Instruction RFB 757, July 25, 2007.
9. *II (Imposto de Importação)* is duty/tax levied on importations into Brazil.

an industrialization process), as long as a certain percentage is incorporated into finished products destined for export. For the remaining percentage (e.g., materials incorporated into finished products that are sold in the Brazilian market), import taxes are only paid when the finished products are sold in Brazil.

First enacted in December 1997, RECOF was designed to increase Brazilian exports and bolster local manufacturing operations in the information technology and telecommunications industries. Over the next several years, the program expanded to the aeronautics (August 2002), automotive (December 2002), and semiconductor (April 2004) sectors. Further legislation detailed RECOF implementation, and subsequent updates in May 2008, August 2008 and May 2009 clarified those efforts.[12] Current participants in the RECOF program include (but are not limited to) Ericsson, Hewlett Packard, Volvo, Samsung and Caterpillar.

[a]   Benefits

The RECOF regime offers significant benefits to Brazilian importers:

- suspension and exemption of II, IPI and PIS/COFINS for eligible imports (for one year, with a possible extension);
- expedited import clearance into in-bond warehouses;
- green lane clearance, which provides express customs clearance within four to eight hours and preferential treatment of imported, export and in-transit products;
- cash flow enhancement by deferring customs duty and import tax payments on domestic sales;
- increase of inventory turnover;
- ability to import products without indicating final destination at the time of import (helps importers to avoid operational issues associated with Drawback[13]);
- minimum bureaucracy in customs clearance; and
- storage cost reduction, both in-house and at the port/airport.

---

10. IPI (*Imposto sobre Productos Industrializados*)— is a Federal Excise Tax applied to manufactured goods.
11. PIS (*Programa de Integração Social (Social Integration Program Contribution)* and COFINS (*Contribuição para o Financiamento da Seguridade Social (Social Security Financing Contribution)*) are levied upon importations as of May, 2004.
12. ADE COANA/ COTEC 01 of May 2008 and ADE COANA/COTEC 02 of August 2008 and ADE COANA/COTEC 02 of May 2009.
13. Drawback generally refers to a refund of customs duties on an importation that is linked to a subsequent exportation (or destruction) of an article.

[2] Requirements for Participation

While each industry has its own specific requirements, there are several common conditions that companies must meet to participate in RECOF. They include:

(a) Industrial operations in Brazil, e.g. assembly, industrialization, modification or reconditioning of raw materials and components.
(b) At least 80% of imported raw materials must be incorporated into the production of industrial goods in Brazil.
(c) Fiscal probity.
(d) Net equity of at least BRL 25,000,000.[14]
(e) Use of specified eight-digit tariff classifications.
(f) Enabled on "Linha Azul," Brazil's Authorized Economic Operator program.[15]
(g) Direct exportation of finished products, or participation in a chain of production which results in export.
(h) Incorporation of at least 50% of the total value of imported goods into exported products, with specific minimum export amounts dictated for specific industries:
   (i) USD 10,000,000 for companies in the information technology, telecommunications, and semi-conductor industries;
   (ii) USD 20,000,000 for companies in any other industry; and
(i) Use a software application such that Brazil's tax authority, Receita Federal do Brasil, can continuously monitor the participant's import and export transactions, including input, output, storage of goods, and taxes due.

[3] RECOF Software

One of the most important requirements for participation in the RECOF program is the use of a software application. The software helps Brazilian companies comply with the regulations that govern RECOF, and helps the Brazilian government control and oversee each company's participation.

The software allows Receita Federal do Brasil to continually access data on each participant's import and export transactions, and it allows participants to expedite the process by which they prepare and present customs documentation. To effectuate all of the necessary controls and to provide real-time information to Receita Federal do Brasil, participants must completely integrate the software into their corporate systems.

---

14. Approximately USD 14,120,000 as of July 2010.
15. "Authorized Economic Operator" refers to a business approved by a customs authority as meeting the requirements of the supply chain security standards of the World Customs Organization, or equivalent local standards. World Customs Organization materials are available at www.wcoomd.org.

Software implementation can be quite involved, and companies considering RECOF must carefully assess the return on investment, which will depend on each company's specific operations. Importantly, the costs of software implementation will vary company to company, depending on the complexity of each company's operations and current IT environment. However, any company enabling RECOF software will have to consider the following costs:

- license for the software;
- implementation costs, including re-designing processes, determining hardware needs, and mapping system integrations;
- allocation of personnel to support RECOF implementation and operations; and
- hardware purchases and infrastructure development.

Automation is the centerpiece of the RECOF program, and often the costs and complexity of software implementation are a determinative factor in a business decision to participate in the program. Because RECOF benefits can be very significant, it is an investment a number of companies will undertake.

§7.03 AUTOMATION NECESSARY TO REALIZE BENEFITS

Despite the experiences reported in Mexico and Brazil with mandated automation, there are very few countries that have actually required automation as a pre-requisite to participation in special programs. Nevertheless, automation procedures for special customs programs is quite commonplace. In some instances, the use of automation appears to be a natural progression of automated business processes. In others, however, the use of technology for a specific customs application occurred quite rapidly as business recognized that it could use technology to realize benefits from special customs programs that while theoretically possible, were not previously practical to achieve. As a result, the realized benefits related to the use of these customs programs expand as technology allows companies to realize benefits that would not be possible without such technological capabilities. If the technology does not create new benefits, it undoubtedly has the potential to reduce cost related to existing benefits. Examples include the US Foreign-Trade Zones Program, European Bonded Warehousing and Processing Trade in China. This section will explore each program, specifically focusing on technology's role in helping companies achieve great savings while maintaining compliant operations.

[A] U.S. Foreign-Trade Zones

[1] Background

A foreign trade zone (FTZ) is an area that is physically located within the United States, but is considered outside of the customs territory of the United States. The United States version of the free port, FTZs are designed to increase the use of American labor and increase capital investment in the United States by allowing activity to occur in the United States prior to the application of US customs laws, thereby equalizing the customs treatment of the activity with similar activities occurring off shore or overseas. Authorized by the Foreign-Trade Zones Act of 1934,[16] foreign trade zones operate as public utilities pursuant to grants from the Foreign Trade Zones Board, an interagency (US Departments of Commerce and Treasury) board established by Congress. US FTZs operate under the supervision of US Customs and Border Protection.

[2] Benefits

A business operating in a foreign trade zone may receive a number of economic benefits:

Duty Deferral: Customs duty and federal excise tax, if applicable, are paid only when merchandise leaves the FTZ and enters the commerce of the United States (not when the merchandise physically arrives in the United States) enhancing the importer's cash flow by delaying the time that duty must be paid.

Duty Elimination on Exports: Merchandise admitted into and re-exported from an FTZ is not subject to duty at all, subject to some limitations on NAFTA transactions.

Duty Reduction (Inverted Tariff Relief): An FTZ user that assembles or manufactures products in a zone may elect to pay duty on an imported component either at the duty rate applicable to the component or at the duty rate applicable to the finished product, resulting in a lower overall duty to the manufacturer. For example, the tariff rate on a complete cellular telephone is zero, while cellular telephone batteries are subject to duty, many at a rate of 3.4%. A US-based cellular phone manufacturer that imports foreign batteries would typically pay a 3.4% customs duty at the time of importation, or about 17 cents of duty on every USD 5 battery. By establishing assembly operations in a zone, however, that manufacturer may bring the battery into the zone, attach it to the body of the cellular phone, and apply the zero rate applicable to the phone. The net result is complete elimination of the customs duty, creating a savings of 17 cents on each USD 5 battery.

Local Property Tax Exemption: The Foreign Trade Zones Act exempts imported inventory and inventory held for export from state and local property taxation.[17] This benefit is particularly valuable in states with high tax rates on inventory, such as Texas. Some states, including Kentucky and Arizona, have state laws that provide additional benefits.

Weekly Entry and Merchandise Processing Fee (MPF) Savings: Normally, every importation into the United States requires the filing of customs entry documentation. In an FTZ, however, no entry is required on arrival to the FTZ (which is outside of customs territory), and all shipments from the FTZ to US locations may be consolidated on a single weekly entry filing. Along with the administrative convenience, a single weekly entry can substantially lower the amount of merchandise processing fee (MPF)

---

16. 19 U.S.C. §§ 81a-81u.
17. 19 U.S.C. § 81(o)(e).

paid by the importer. MPF is a US government "customs user fee" charged to importers in additional to customs duty. The rate is 0.21% of value, but the MPF is subject to a cap of USD 485 per customs entry. An FTZ user pays just one MPF fee per week, a maximum of USD 25,220 in a 52-week year. The associated savings can be significant; an importer which would normally file 25 entries each week at the capped amount can save over USD 600,000 annually in MPF.

Logistical Benefits: With approval from the local US Customs Port Director, FTZ operators utilizing Direct Delivery procedures receive merchandise directly to their facilities, eliminating delays from processing back-up at the port of arrival. Direct Delivery can be especially valuable to businesses utilizing a "just in time" inventory management system. Also, implementing certain inventory control and recordkeeping systems (ICRS) may improve a business' ability to track merchandise more accurately.

[3]   *Inventory Control and Recordkeeping System*

US Customs regulations require that the FTZ operator maintain an ICRS capable of accounting for all merchandise admitted to or removed from the zone, produce accurate and timely reports, identifying any inventory shortages and overages.[18] The ICRS is key to effective FTZ utilization. There are no requirements for automation of the ICRS. Automation has clearly enabled growth of the FTZ program, however. As companies have increased volume of transactions, and emphasized supply chain speed with just-in-time inventories, cross-dock distribution operations, and vendor and third-party logistic provider partnerships, FTZ accounting software has become more robust and pervasive. Virtually all significant FTZ operations are automated to some extent. In one industry sector, there is a direct connection between growth and automation.

[4]   *Petroleum Refining FTZs*

The industry with the most participation in the FTZ program (by value of foreign merchandise) is oil refining. For government fiscal year 2008, foreign-status crude oil and petroleum products received in US foreign trade subzones totaled more than USD 229 billion.[19] The primary benefit to oil refiners is the opportunity to avoid an inverted tariff. Many of the outputs of the refining process have a lower duty rate than that of the raw material: crude oil. Crude oil with an API gravity of at least 25 degrees carries a specific duty rate of US 10.5 cents per barrel. While some primary refinery outputs, such as gasoline and jet fuel, have duty rates that exceed the crude oil rate, other outputs, such as residual fuel have a lower duty rate of US 5.25 cents per barrel. Refinery process by-products such as petroleum coke, sulfur, refinery gases and asphalt have no duty at all. In concept, a FTZ refining operation should allow the

---

18. 19 C.F.R. Part 146, Subpart B.
19. 70th Annual Report of the Foreign-Trade Zones Board to the Congress of the United States. December 2008. http://ia.ita.doc.gov/ftzpage/annualreport/ar-2008.pdf. Foreign status merchandise is imported merchandise which has not yet been released from custom's custody.

refiner to elect the lower of the duty rates applicable to a specific finished product or the crude oil.

While it is easy to understand the conceptual application of the FTZ program to allow oil refiners to save customs duties, practical application is far more difficult. FTZ manufacturing rules were designed for product assembly, taking a variety of component parts and assembling them into a finished product. With oil refining a large number of finished products are derived from one input, crude oil (and in many instances a few additional feedstocks), subjected to a continuous process. Moreover, the duty rate is determined by volume, a specific rate of duty assessed on each barrel, yet because of chemical changes caused by the refining process the volume of outputs from an oil refinery exceeds the volume of inputs.

To access duty savings from FTZs, oil refiners needed two items: regulatory definition of a standard accounting methodology, and automation. These two items did not occur for more than 40 years after the FTZ manufacturing provisions were first enacted.

In 1992, with industry participation, US Customs adopted new regulations for petroleum refineries in FTZs designed to establish a reasonable accounting procedure consistent with the Foreign Trade Zone Act.[20] Most notably, the new regulations established a new method of FTZ accounting called producibility, which permitted the FTZ operator to attribute refined products to feedstocks not based on what was produced, but rather on "what could have been produced." The regulations required that all final products removed from or consumed within a petroleum refinery subzone be attributed to feedstocks which were actually admitted to the FTZ, but the specific attribution is left up to the refiner based on producibility. Under producibility, attribution of final products is allowable to the extent that the quantity of such products *could* have been produced from the available feedstocks, using the industry standards of potential production on a practical operating basis, as published in a schedule, T.D. 66-16.[21]

While the approval of producibility was a huge breakthrough, it was only part of what was needed to actually recognize FTZ savings. Not only was it necessary to account for producibility in a FTZ ICRS system, but to actually optimize FTZ use refiners had to predict optimal producibility with each shipment received at the FTZ refinery. Variables include the types of products that will be made, finished product exports, the types of feedstocks used in refining and the availability of preferential duty rates for the feedstocks (such as NAFTA qualification).

Simultaneously with the effort to amend the FTZ regulations to accommodate oil refineries, one oil company, Amoco, and one accounting firm, Arthur Andersen, undertook a joint project to develop FTZ refinery software. The software was introduced in 1995, and within 5 years more than 60 oil refineries began operating as FTZs. That software, today known as PetroZone,®[22] remains the dominant FTZ software in the refining sector.

---

20. 19 C.F.R. Part 146, Subpart H, Petroleum Refineries in Foreign-Trade Subzones.
21. 19 C.F.R § 146.95(a).
22. PetroZone is licensed by EYPS, an affiliate of Ernst & Young, LLP.

PetroZone is a comprehensive FTZ accounting system, generating all necessary forms (paper and electronic for interfacing with Customs' electronic systems) required by US Customs for foreign feedstock admissions to the FTZ (CBPF 214), performing the weekly attribution process, managing in-bond transportation of products and generating weekly customs entries (CBPF 7501-06) for products produced at the FTZ and entered into the US commerce. PetroZone logic has been designed to process the refinery data in mass, providing a solution for duty tracking caused by the volumetric gain. The automated system also contains numerous validations that are specific to the petroleum and petrochemical industries, such as checking the proper API gravity class when applying the producibility percentages or tariff classification, checking that the volume-to-weight ratio is reasonable for each transaction, and ensuring that attributions of brand-to-brand, or feedstock-to-feedstock shipments meet the "adjusted basis" requirements of US Customs' Regulatory Audit Division.

The impact of the software is so significant in FTZ refinery operations that US Customs determined that it needed to develop specific audit procedures based on a refiner's specific use of the software. The basic procedure was first developed as part of the FTZ audit of Mobil Oil Company in 1996, the first comprehensive audit of a refinery FTZ utilizing producibility accounting. Today, all refinery FTZ audits begin with standard systems queries of PetroZone, or another automated system. US Customs oil refinery FTZ audit teams include a systems specialist who determines the queries and evaluates the data generated.

Effectively, it took both the regulatory change (which permitted producibility) and the development of appropriate automation software for oil refineries to use the FTZ program. The regulations did not mandate that FTZ refineries use automation software to attribute finished products to feedstocks, but it was a practical necessity. As a result, enforcement also became software focused. Quite simply, US Customs had to understand the software in order to effectively audit the FTZ operation. An entire industry sector (in fact, the largest industry sector) effectively uses the FTZ program because of enabling automation, and US Customs' enforcement program for this sector has become uniquely focused on auditing the automated system.

[B]   Customs Warehousing in the European Union

[1]   Overview and Benefits

The European Customs Warehouse program[23] allows importers of non-Community goods (i.e., goods imported from outside the EU) into the EU to store the goods without paying customs duties, import VAT and excise duty (where appropriate) until goods are shipped to an EU customer and leave the warehouse. If the goods are destined to a customer outside the EU, the goods can be stored in transit and transported in-bond with the benefit of duties and VAT being paid only in the country of destination. This

---

23. http://ec.europa.eu/taxation_customs/customs/procedural_aspects/imports/customs_warehouses/index_en.htm.

duty deferral benefit allows companies utilizing the program a cash-flow management option for relief of duty and VAT payments. Additional benefits of Customs Warehousing include having an unlimited storage period for goods, periodic electronic filing options where applicable, no import duty risks on obsolete stock, and minimal interference from the customs authorities. There are both public and private customs warehouses.[24] The Type E private customs warehouses do not require a physical presence authorization for the storage of goods. Effectively, the user's recordkeeping and accounting systems receive the authorization rather than a defined physical location. As such, Type E warehouses require the highest level of administrative organization and recordkeeping procedures to track the flow of merchandise. Many companies outsource this type of automated recordkeeping procedures, as well as other bonded warehouse operations to various types of logistics service providers.

[2]   Focus on Automation

To effectively manage Customs Warehousing operations, automation software has been important. A challenge for many companies and software providers in the EU is the lack of consistency and coordination between the various customs authorities within the EU. Consequently, different software providers operate within different jurisdictions. Some countries within the EU offer automated systems for different types of electronic filing for customs purposes. In the UK for example, for accelerated processing and electronic reporting, companies may apply for Customs Freight Simplified Procedures (CFSP) to import goods under warehousing procedures.

[3]   Customs Freight Simplified Procedures (CFSP)[25]

The CFSP is an electronic customs system administered by HM Revenue & Customs (HMRC) for importation processing. The system is designed to accelerate the release of imported goods from Customs by providing importers (including customs warehouse keepers) the opportunity to submit frontier declarations (including National Transit[26]) and all supplementary declarations by electronic means. It also helps importers realize certain cash-flow benefits through local clearance and temporary storage procedures. Through the electronic interface, HMRC has greater visibility of customs activity, enabling them to more accurately target "high-risk" importers and build partnerships with compliant, "low-risk" importers. The electronic interface also enables warehouse

---

24. Public warehouses (types A, B, F) can be used by any 'depositor' for storing goods. Private warehouses (types C, D, E) are restricted to the warehousekeeper. Type F warehouses are operated directly by the customs authorities.
25. Available at: http://customs.hmrc.gov.uk/channelsPortalWebApp/channelsPortalWebApp.portal?_nfpb=true&_pageLabel=pageImport_InfoGuides&propertyType=document&id=HMCE_PROD_010383.
26. National Transit is used to move goods which are imported into the UK but have not yet cleared customs. Such goods can be moved from the UK frontier to an approved inland location where they can be released to free circulation using a simplified procedure, provided a national transit declaration has been made at the UK frontier.

users to use third-party service providers to submit electronic declarations. With these technological capabilities and electronic interfaces, European Customs Warehouses can effectively operate all day long, 365 days per year. As in the case for certain US FTZs, for large distribution facilities operating as European customs warehouses, the ability to electronically track all inventory movement (receipts and shipments) allows the warehouse keeper to manage high volume activity and reap the associated tax benefits. These savings would not be achievable without automation technology.

[C] Processing Trade in China

[1] Overview and History

Processing Trade in China ("PTC") is a bonded manufacturing system that allows a manufacturer to import raw materials used for the production of export goods without the payment of duty or Value Added Tax ("VAT").[27] There are two types of PTC operations: contract manufacturing and toll manufacturing. Contract manufacturing is a buy-sell arrangement for the finished product; the processor is required to purchase imported raw materials, and the processor is qualified to claim an export VAT refund. Toll manufacturing is a fee for services arrangement where raw materials are mostly provided by a foreign principal on a free-of charge-basis, and the processor is paid for assembling the materials. Toll manufacturers cannot recover VAT on export.

PTC operations are a significant part of import and export trade in China. Recently published figures indicate that almost 50% of China's entire trade occurs under PTC.[28] Primary products exported under PTC include electronic goods, machinery, clothing, shoes, suitcases, and toys.

[2] Primary Requirement for Participation in PTC: Customs Handbook

The primary requirement for participation in PTC is the use of a Customs Handbook, the People's Republic of China Customs' primary administrative tool to track the overall movement of imported inputs and exported outputs under PTC.[29] Each Handbook tracks the specific details of imports and exports, including the imported raw material inputs used to manufacture the finished product for export, the supplier, the port of importation, and input consumption. Bills of materials ("BOM") are the primary means by which China Customs determines which and how much imported raw material inputs were used to manufacture the one unit finished good(s) for export. Manufacturers provide the BOM to China Customs before importation of the raw

---

27. Measures of Customs of the People's Republic of China on the Supervision of Processing Trade Goods (Revised in 2008). Order of the General Administration of Customs [2008] No.168. January 14, 2008.
28. 2009 Annual Report: China Customs. Available at: http://english.customs.gov.cn.
29. Please note that there may be provincial or regional differences in actual implementation, timing and practice of the Handbook.

material inputs. At export China Customs makes deductions from the Handbook[30] based on the bill of material consumption rate and reflects these deductions on the export declaration. Below is a diagram that shows these steps and relationships.

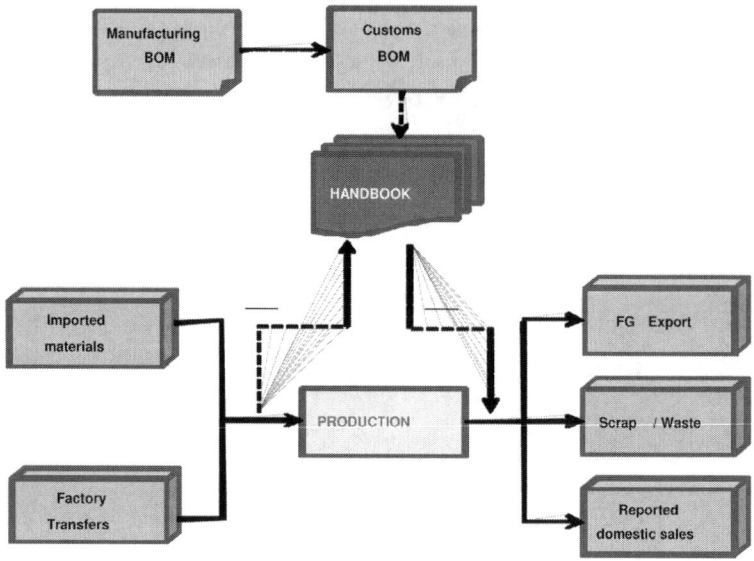

At the end of the period encompassed by the Handbook (usually six months or one year), the manufacturer works with China Customs to reconcile the Handbook, account for discrepancies and as necessary pay duty/VAT on shortages. In theory, the quantities of bonded raw materials in the Handbook, the manufacturer's ERP system, and its physical inventory should match and balance. In practice, however, discrepancies are quite common. Discrepancies may be attributable, for example, to a transfer between bonded and non-bonded raw material inventory, inclusion of domestic sales with bonded raw materials, or failure to adequately account for production changes not reflected on standard bills of materials. There may be serious consequences if a manufacturer is unable to reconcile these discrepancies. China Customs may: assess additional customs duty or financial penalty based on the amount of underpaid duties or goods value; suspend bonded manufacturing; suspend the issuance of a new Handbook; perform more intensive goods and documentation inspection; or perform more frequent post-importation audits.[31]

---

30. Measures of Customs of the People's Republic of China on the Supervision of Processing Trade Goods (Revised in 2008). Order of the General Administration of Customs [2008] No.168. January 14, 2008.
31. Regulations of the People's Republic of China on Implementing Customs Administrative Penalty, promulgated by Decree No. 420 of the State Council of the People's Republic of China on September 19, 2004, and effective as of November 1, 2004.

[3] Development of the Electronic Customs Handbook

Historically, all Handbooks were paper-based, contract-specific, and limited to one BOM. These requirements made it difficult to accurately update and maintain the Handbooks, as many manufacturers exported multiple products that used the same set of materials and were unable to properly calculate the bill of material consumption rate. As a result, many PTC participants were unable to balance the quantities of bonded raw materials in the Handbook, the company's ERP system, and its physical inventory— and account for discrepancies. Non-compliance was very common, as were penalty assessments.

Recognizing that both industry and government operations were made inefficient by Handbook inaccuracy, in 2006 China Customs developed an Electronic Customs Handbook ("e-Handbook") option.[32] Three types of e-Handbooks have been introduced. The first type uses an interface, through which the manufacturer sends China Customs details about its BOMs and import/export transactions on Excel spreadsheets. The second type uses a software system that electronically transmits data from the manufacturer to China Customs. This typically requires the manufacturer to manually input data from its ERP system into the software system for transmission to China Customs. The third type is an online system through which the manufacturer links its ERP system directly to China Customs to facilitate real-time data access.

China Customs' e-Handbook system is a significant improvement from the paper-based Handbook system. Each e-Handbook type improves the tracking of imports and exports by more frequently performing data file transfers from the participant's ERP system to China Customs' system. The e-Handbook system also reduces the number of Handbooks that a manufacturer must have, as PTC participants now receive an e-account that serves as one, company-wide Handbook, instead of contract-specific Handbooks. Another improvement is BOM consolidation. Unlike the paper-based system, each BOM is pooled to one e-Handbook balance, making it easier for manufacturers to approximate the BOM and reconcile.

The development of the e-Handbook has improved how manufacturers report and manage their PTC operations. It has reduced risk, and lowered costs, for many manufacturers. Interestingly, while the third type of e-Handbook would appear to have provided the greatest degree of automation and risk mitigation, anecdotal evidence suggests that few manufacturers have selected it. Concerns include cost, concerns over the level of government intrusion, and implementation time and effort. This presents an interesting contrast to the RECOF program in Brazil, in which a directly linked system is a pre-requisite to benefits. In China, where this type of system is not required, it is not commonly used, despite a higher degree of accuracy.

---

32. Measures of the Customs of the PRC for Networked Regulation of Processing Trade Enterprises. Order of the General Administration of Customs [2006] No. 150. June 14, 2006.

§7.04 USE OF AUTOMATED TOOLS

Country of origin preference programs are among the most valuable duty reduction tools available to importers, yet are routinely viewed as suboptimized. Origin determination is not harmonized. Each country has its own general standard for origin, a "non-preferential" origin rule that is used to generally determine where a product is considered to have originated. In the US, for example, a manufactured product is considered to originate in the last location of "substantial transformation," a subjective standard that evaluates the change in a product's name, character, or use resulting from a processing or assembly operation.[33]

"Preferential" origin refers to the origin of a product according to rules of origin set forth in a particular trade agreement or preference program. Items qualifying as a product of a "preferred" trading partner receive a reduced rate of duty. There are hundreds of these types of programs around the world, and many of them have very detailed, products specific rules.[34] Moreover, the rules can follow very different patterns: some look at the value added in the preferred country or region, some look at the shift in tariff classification of categories of materials from outside the preferential area, and others define specific processes which control origin.

This scenario is quite complex for both importers and exporters. Rules of origin are rarely harmonized even within a country. For example, a US importer has to consider non-preferential rules of origin for every product imported, and can often benefit from considering a variety of preferential rules of origin, such as those under the North American Free Trade Agreement (NAFTA), the US-Australia Free Trade Agreement, the Generalized System of Preferences, or the African Growth and Opportunity Act Preference Program. Products made by a Mexican manufacturer, on the other hand, can avoid duties in places as diverse as the US, European Union, and Japan, but to do so must qualify products under different rules of origin for each free trade agreement. And, of course, to optimize opportunity, the manufacturer must be able to do this predictably— knowing that duty must be incurred or can be avoided before a sale is made.

Consequences of origin determination go far beyond determining eligibility for preferential duty rates. Origin can also impact the potential applicability of antidumping or countervailing duties, the applicability of special trade measures such as safeguards or retaliatory actions, admissibility of a product into a country, the type of labeling that must accompany a product, and eligibility of products for sales to governments. Perhaps because the variables that impact the optimization of country of origin process are very product specific, comprehensive solutions along the lines of those used to facilitate special programs have been much difficult to develop. Effective

---

33. The substantial transformation standard developed as a judicial doctrine, see Anheuser-Busch Brewing Association v. United States, 207 U.S. 556 (1908); United States v. Gibson-Thompsen Co., 27 C.C.P.A. 267 (1940); 19 CFR 134.35(a) and has been incorporated into regulation, 19 C.F.R. §134.1(b).
34. The World Trade Organization maintains a listing of regional trade agreements at http://rtais.wto.org.

automated solutions include tools to assist with origin determination, origin data integrity, and scenario planning, as well as databases to maintain origin information.[35]

Origin determination tools are, in effect, automated decision trees. The tools access costed bill of material information which also contains component country of origin data. This combination of data can be analyzed to determine qualification under a variety of preference program. In some cases, these tools are standalone, regularly managing the determination of origin under the variety of rules sets that may be applicable. In others, they are integrated with production systems, evaluating information which can be fed back to production planners to optimize preference qualification, or to continually use production systems to verify origin qualification.

Tools can also be important for tracking origin. As products move from a raw materials inventory to work in process, it is easy for the origin of the products to be lost. Properly tracking origin through the entire process is often times critical to documenting eligibility for preference programs. Tools which can identify the critical points in production process for origin determination can often eliminate the need for detailed component analysis, saving time and money. Many automated systems have portals for tracking origin information once determined, so that the results of an analysis can be readily accessed throughout an organization.

Finally, as mentioned earlier, predictively assessing origin, and understanding the origin determination consequences of potential changes in sourcing or production process, is a leading practice. Procurement departments often consider preferential origin in assessing the landed cost of material purchases. It is more complicated, however, to assess the importance of origin of a particular component material in the qualification of a finished product for one or more third country destinations. Matching the input and output origin requirements and comparative costs can be extremely beneficial in optimizing the use of tariff preference programs. Tools that assist with scenario planning, looking at the consequences of "what if" procurement decisions on materials and possible end destinations of finished products have proven very beneficial to global traders.

§7.05 CONCLUSION

Customs automation is often thought of as compliance focused, keeping necessary data for customs declarations and facilitating electronic filing and recordkeeping. Automation, however, has also played a big role in tariff reduction planning— tax planning for customs duties— in the two areas most critical to global traders, special programs and trade preferences. Some of this automation is government mandated, essentially an entry fee for accessing tax savings. More frequently, automation enables the tax

---

35. County of origin and trade preference management applications are developing rapidly. The SAP Business Objects Global Trade Services application for example includes a trade preference management function, http://www.sap.com/solutions/sapbusinessobjects/large/governance-risk-compliance/globaltradeservices/index.epx, as do applications offered by MIC, http://www.mic-cust.com/software-solutions/origin-calculation-supplier-solicitation, and TradeBeam, http://tradebeam.com/solutions/compliance/export_management/.

savings, whether in the form of a comprehensive automated accounting system for a special program, or as a tool allowing rapid assessment and trade preference scenario planning. As the technology becomes more pervasive, it also becomes the framework for government enforcement. Looking ahead, as business looks to squeeze additional costs out of the supply chain using special customs programs and trade preferences, and as government enforcement focuses on the compliance requirements for special program and trade preference eligibility, it seems likely that automated systems will play an even more critical role in assisting business to achieve savings and meet compliance obligations.

§7.06   REFERENCES

China Customs, *2009 Annual Report*. Available at: http://english.customs.gov.cn.

Decreto para el Fomento de la Industria Manufacturera, Maquiladora y de Servicios de Exportacion (Manufacturing Industry, Maquiladora, and Exportation Services Program), *Diario Oficial de la Federación* [D.O.] 1 de noviembre de 2006 (Mex.).

Foreign Trade Zones Board, *70th Annual Report of the Foreign-Trade Zones Board to the Congress of the United States*. December 2008. http://ia.ita.doc.gov/ftzpage/annualreport/ar-2008.pdf.

General Administration of Customs, *Measures of the Customs of the PRC for Networked Regulation of Processing Trade Enterprises*, Order No.150. June 14, 2006.

General Administration of Customs, *Measures of Customs of the People's Republic of China on the Supervision of Processing Trade Goods*, Order No.168. January 14, 2008.

HMRC, *Customs Freight Simplified Procedures*. Available at: http://customs.hmrc.gov.uk/channelsPortalWebApp/channelsPortalWebApp.portal?_nfpb=true&_pageLabel=pageImport_InfoGuides&propertyType=document&id=HMCE_PROD_010383.

Integration Point, Inc., *Leverage Mexican and US Trade Opportunities with Integration Point Mexican Trade* (2009). http://www.integrationpoint.com/products/brochures/IP_MexicanTrade_ProductBrochure_2009.pdf.

State Council of the People's Republic of China , *Regulations of the People's Republic of China on Implementing Customs Administrative Penalty*, Decree No. 420 of September 19, 2004, and effective as of November 1, 2004.

SERIES ON INTERNATIONAL TAXATION

1. Alberto Xavier, *The Taxation of Foreign Investment in Brazil*, 1980 (ISBN 90-200-0582-0).
2. Hugh J. Ault & Albert J. Radler, *The German Corporation Tax Law with 1980 Amendments*, 1981 (ISBN 90-200-0642-8).
3. Paul R. McDanie! & Hugh J. Ault, *Introduction to United States International Taxation*, 1981 (ISBN 90-6544-004-6).
4. Albert J. Radler, *German Transfer Pricing/Prix de Transfer en Allemagne*, 1984 (ISBN 90-6544-143-3).
5. Paul R. McDanie! & Stanley S. Surrey, *International Aspects of Tax Expenditures: A Comparative Study*, 1985 (ISBN 90-654-4163-8).
6. Kees van Raad, *Nondiscrimination in International Tax Law*, 1986 (ISBN 906544-266-9).
7. Sijbren Cnossen (ed.), *Tax Coordination in the European Community*, 1987 (ISBN 90-6544-272-3).
8. Ben Terra, *Sales Taxation. The Case of Value Added Tax in the European Community*, 1989 (ISBN 90-6544-381-9).
9. Rutse! S.J. Martha, *The Jurisdiction to Tax in International Law: Theory and Practice of Legislative Fiscal Jurisdiction*, 1989 (ISBN 90-654-4416-5).
10. Pau! R. McDanie! & Hugh J. Au!t, *Introduction to United States International Taxation* (3rd revised edition), 1989 (ISBN 90-6544-423-8).
11. Manue! Pires, *International Juridicial Double Taxation of Income*, 1989 (ISBN 90-6544-426-2).
12. A.H.M. Daniels, *Issues in International Partnership Taxation*, 1991 (ISBN 90-654-4577-3).
13. Arvid A. Skaar, *Permanent Establishment: Erosion of a Tax Treaty Principle*, (ISBN 90-6544-594-3).
14. Cyrille David & Geerten Michielse (eds), *Tax Treatment of Financial Instruments*, 1996 (ISBN 90-654-4666-4).
15. Herbert H. Alpert & Kees van Raad (eds), *Essays on International Taxation*, (ISBN 90-654-4781-4).
16. Wolfgang Gassner, Michael Lang & Eduard Lechner (eds), *Tax Treaties and EC Law*, 1997 (ISBN 90-411-0680-4).
17. Gloria Teixeira, *Taxing Corporate Profits in the EU*, 1997 (ISBN 90-4110703-7).
18. Michael Lang et al. (eds), *Multilateral Tax Treaties*, 1998 (ISBN 90-4110704-5).
19. Stef van Weeghel, *The Improper Use of Tax Treaties*, 1998 (ISBN 90-4110737-1).
20. Klaus Vogel (ed.), *Interpretation of Tax Law and Treaties and Transfer Pricing in Japan and Germany*, 1998 (ISBN 90-411-9655-2).
21. Bertil Wiman (ed.), *International Studies in Taxation: Law and Economics; Liber Amicorum LeifMuten*, 1999 (ISBN 90-411-9692-7).
22. Alfonso J. Martin Jimenez, *Towards Corporate Tax Harmonization in the European Community*, 1999 (ISBN 90-411-9690-0).

23. Ramon J. Jeffery, *The Impact of State Sovereignty on Global Trade and International Taxation*, 1999 (ISBN 90-411-9703-6).
24. A.J. Easson, *Taxation of Foreign Direct Investment*, 1999 (ISBN 90-4119741-9).
25. Marjaana Helminen, *The Dividend Concept in International Tax Law: Dividend Payments Between Corporate Entities*, 1999 (ISBN 90-411-9765-6).
26. Paul Kirchhof, Moris Lehner, Kees van Raad, Arndt Raupach & Michael-Rodi (eds), *International and Comparative Taxation: Essays in Honour of Klaus Vogel*, 2002 (ISBN 90-411-9841-5).
27. Krister Andersson, Peter Melz & Christer Silfverberg (eds), *Liber Amicorum Sven-Olof Lodin*, 2001 (ISBN 90-411-9850-4).
28. Juan Martin Jovanovich, *Customs Valuation and Transfer Pricing: Is It Possible to Harmonize Customs and Tax Rules?*, 2002 (ISBN 90-411-9888-1).
29. Stefano Simontacchi, *Taxation of Capital Gains under the OECD Model Convention: With Special Regard to Immovable Property*, 2007 (ISBN 97890-411-2549-1).
30. Michael Lang, Josef Schuch, & Claus Staringer (eds), *Tax Treaty Law and EC Law*, 2007 (ISBN 978-90-411-2629-0).
31. Duncan Bentley, *Taxpayers' Rights: Theory Origin and Implementation*, 2007 (ISBN 978-90-411-2650-4).
32. Sergio Andre Rocha, *Interpretation of Double Taxation Conventions: General Theory and Brazilian Perspective*, 2008 (ISBN 978-90-411-2822-5).
33. Robert F. van Brederode, *Systems of General Sales Taxation: Theory, Policy and Practice*, 2009 (ISBN 978-90-411-2832-4).
34. John G. Head & Richard Krever (eds), *Tax Reform in the 21st Century: A Volume in Memory of Richard Musgrave*, 2009 (ISBN 978-90-411-2829-4).
35. Jens Wittendorff, *Transfer Pricing and the Arm's Length Principle in International Tax Law*, 2010 (ISBN 978-90-411-3270-3).
36. Marjaana Helminen, *The International Tax Law Concept of Dividend*, 2010 (ISBN 978-90-411-3206-2).
37. Robert F. van Brederode (ed.), *Immovable Property under VAT: A Comparative Global Analysis*, 2011 (ISBN 978-90-411-3126-3).
38. Dennis Weber & Stef van Weeghel, *The 2010 OECD Updates: Model Tax Convention & Transfer Pricing Guidelines — A Critical Review*, 2011 (ISBN 978-90-411-3812-5).
39. Yariv Brauner & Martin James Mcmahon, Jr. (eds), *The Proper Tax Base: Structural Fairness from an International and Comparative Perspective— Essays in Honour of Paul McDaniel*, 2012 (ISBN 978-90-411-3286-4).
40. Robert F. van Brederode (ed.), *Science, Technology and Taxation*, 2012 (ISBN 978-90-411-3125-6).